The Facts On File

DICTIONARY
OF
TELEVISION,
CABLE, AND VIDEO

The Facts On File

DICTIONARY

OF

TELEVISION,

CABLE, AND VIDEO

ROBERT M. REED

and

MAXINE K. REED

Facts On File®

AN INFOBASE HOLDINGS COMPANY

The Facts On File Dictionary of Television, Cable, and Video

Facts On File, Inc.
460 Park Avenue South
New York NY 10016

Library of Congress Cataloging-in-Publication Data
Reed, Robert M.
　　The Facts On File dictionary of television, cable, and video / Robert M. Reed and Maxine K. Reed.
　　　　p.　cm.
　　ISBN 0-8160-2947-4 (alk. paper)
　　1. Television broadcasting—Dictionaries　2. Cable television—Dictionaries. 3. Video recordings—Dictionaries.　I. Reed, Maxine K.　II. Title.　III. Title: Dictionary of television, cable, and video.
　　PN1992.18.R44　1994
　　384.55′03—dc20　　　　　　　　　　　　　　　　　　　　94-1221

A British CIP catalogue record for this book is available from the British Library.

Facts On File books are available at special discounts when purchased in bulk quantities for businesses, associations, institutions or sales promotions. Please call our Special Sales Department in New York at 212/683-2244 or 800/322-8755.

Jacket design by Linda Kosarin

Printed in the United States of America

MP VC 10 9 8 7 6 5 4 3 2 1

This book is printed on acid-free paper.

CONTENTS

INTRODUCTION

The Premise

This dictionary contains words and terms from three overlapping fields of electronic media. The fields—television, cable, and video—are closely intertwined today in research and teaching and in professional practices. Communication departments in colleges teach all three disciplines, and it is hoped that a single volume covering the fields will be a valuable resource for students, researchers, and professionals in communications, as well as for other interested persons.

The Fields

Although television, cable, and video use different technology to reach an audience, they have the same objective—communication. They also share the same production techniques and engineering equipment, and they borrow extensively from one another in the process of distribution.

The language of the older form of broadcast television has been adapted and modified by the newer media of cable and video. New definitions or twists on existing terms have become industry jargon and are used interchangeably.

This dictionary covers terms in ten specific areas as they relate to the electronic media:

Advertising
Agencies/Associations/Companies/ Unions
Broadcasting/Cablecasting
Educational/Corporate Communications
Engineering
General Terms and Processes
Government/Legal
Home Video
Production
Programming

Unless a word or term is directly related to both film and the electronic media, it is not included in this dictionary. Film and the electronic media are two distinct disciplines and are usually taught in two different departments at universities. Their practitioners have different perspectives and vocabularies. This is a dictionary of the electronic media.

How to Use This Dictionary

The words, acronyms, abbreviations, phrases and terms related to the ten areas are defined. Some of the entries are short while others present the definitions with some background and are therefore necessarily longer. The terms are arranged in a modified A-to-Z alphabetical order on the word-by-word principle. There are no entries under X and Y.

This dictionary is extensively cross-referenced. When a term appears in a definition in small capitals, it signals the user that it is discussed in another definition and can be found at the appropriate alphabetical location. Some entries are not followed by a definition, but the reader is directed to a discussion of the term with a "See . . ." reference. The term "See also . . ." is also found at the end of some definitions to direct the user to a related entry. All three types of cross-references are identified by small capitals. Such references occasionally use a related word to direct the reader to the reference (e.g., "broadcast" rather than "broadcasting"). A pronunciation guide is included in a definition only if appropriate.

Apologia

In the process of creating this dictionary, there have most certainly been sins of omission and commission. The nagging questions of what should and should not be included had to be addressed, for there never seems to be enough time or space. In addition, the three fields are fast moving and everchanging. By its very nature, some of the information may be dated by the time the book reaches the reader. The definitions are considered accurate as of 1993.

This dictionary is an interpretation of the past and the present—and it provides a glimpse into the future. The authors hope it will be a useful addition to the literature. Suggestions for additions, deletions, or changes for future editions are most welcome.

RMR and MKR—1994
Reed-Gordon Books
285 Burr Road
East Northport
New York 11731

A

A. C. Nielsen Company a subsidiary of Dun and Bradstreet known for its RATINGS service for broadcast and cable television. Audience measurements by Nielsen Media Research Divisions are the commonly accepted index to the popularity of a nationally distributed television program in the United States and Canada. (See also ABCD COUNTIES, AGB TELEVISION RESEARCH, and SHARE.)

"A" title a recent major motion picture that attracted a sizeable audience in its theatrical run and has now been released for home video. It was produced by one of the major Hollywood studios for more than $10 million and had box office revenues of at least $25 million. (See also "B" TITLE.)

AB roll editing a complex videotape editing method that makes full use of computerized editing controllers and SOCIETY OF MOTION PICTURE AND TELEVISION ENGINEERS (SMPTE) Time Code capabilities. The AB roll method permits dissolves, wipes, superimpositions, and other special effects between scenes and sequences.

One scene on a prerecorded tape from machine A and another scene from a tape on machine B are rolled (played back) simultaneously, along with a blank tape on a third machine. The computer determines the edit points on A and B, and the result ends up on the third tape. The AB roll technique is usually used with ONLINE EDITING. (See also RIPPLE EFFECT.)

AB switch a switch that is used to conveniently select between two electronic input signals. It enables the cable input to the television set to be bypassed and the reception switched to the incoming signal from a broadcast receiving ANTENNA on the roof.

ABC Television Network one of the three major, commercial, full-service, national television networks, headquartered in New York City. It was purchased by a smaller GROUP BROADCASTER, Capital Cities Communications, for $3.5 billion in 1986 and renamed Capital Cities/ABC Inc., but the network is still known by the ABC Television Network name.

In addition to the network of some 230 affiliated stations, the parent company owns and operates eight individual affiliated stations (O & Os).

ABCD counties the system devised by the A. C. NIELSEN COMPANY whereby counties in the United States are designated one of four sizes (A, B, C, or D) of geopolitical entities. The designations are based on the number of people residing in a given county and its proximity to a metropolitan city.

An "A" county is the largest and surrounds or is near one of the 25 largest U.S. cities; a "B" county has a population of more than 150,000; a "C" county has a population of more than 35,000; and all of the smaller counties are designated as "D" counties. (See also DESIGNATED MARKET AREA, MARKET, and METRO AREA.)

above-the-line costs   one part of a method of financial accounting for production. The term refers to the placement of the costs for the creative elements of a show on an accounting sheet. Charges associated with the performers, writers, or producer, and with the artwork or design, are placed in this category. These costs are placed in columns literally above a line on an accounting sheet, to distinguish them from BELOW-THE-LINE COSTS.

Academy of Canadian Cinema and Television   a nonprofit organization that awards the Canadian equivalent of the OSCAR AWARDS and EMMY AWARDS. The group presents the Gemini (English Canadian) and Gemeaux (French Canadian) awards for excellence in television each year in various categories from Best Actor to Best Director. (See also BRITISH ACADEMY OF FILM AND TELEVISION ARTS [BAFTA].)

Academy of Family Films and Family Television (AFFFT)   a Los Angeles-based nonprofit organization that consists of individuals who encourage family entertainment.

Academy of Motion Picture Arts and Sciences (AMPAS)   an honorary membership organization composed of outstanding individuals in the motion picture industry. The academy provides information services and presents the annual OSCAR AWARDS.

Academy of Television Arts and Sciences (ATAS)   a nonprofit organization responsible for the presentation of the annual EMMY AWARDS for nighttime PRIME-TIME programming each September. ATAS was formed in 1978 by the dissident Hollywood chapter of the NATIONAL ACADEMY OF TELEVISION ARTS AND SCIENCES (NATAS).

access   See PRIME-TIME ACCESS.

access channels   See CUPU LEASED ACCESS CHANNELS and PEG CHANNELS.

account   a colloquial term referring to the business relationship between two companies. The firm that services the other is said to have that company as an account.

Accrediting Council on Education in Journalism and Mass Communications (ACEJMC)   an organization that consists of associations related to journalism education. It accredits academic programs in more than 200 academic sequences (including Radio–TV) in nearly 100 colleges and universities in the United States.

Accuracy in Media (AIM)   a Washington D.C.-based, nonprofit organization that monitors the media. AIM is known largely for its study of television news bias. Its criticism of television is often from a conservative perspective, in contrast to the analysis and views of FAIRNESS AND ACCURACY IN REPORTING (FAIR).

ACE awards   annual awards, in the form of statuettes, honoring excellence and achievement in made-for-cable programming. They are given in some 82 categories, ranging from Best Movie or Miniseries to Best Stand-up Comedy Special, by the NATIONAL ACADEMY OF CABLE PROGRAMMING.

ACORN   a statistical analysis research methodology. The acronym stands for "A Classification of Residential Neighborhoods." The system classifies all U.S. and U.K. households into DEMOGRAPHIC segments based on some part of a household address, such as the ZIP code. The system is sometimes used to identify audiences for cable and television programming for SPOT (COMMERCIAL TIME) sales, but it is more often used by ADVERTISING AGENCIES to target direct mail campaigns, including promotions for home video SPECIAL INTEREST (SI) PROGRAMMING. The system is somewhat similar to the CLUSTER ANALYSIS system called PRIZM.

across the board   a type of program scheduling strategy in which individual programs in a series are scheduled at the same time each weekday. The programs are said to run "across the board" and the technique is designed to foster tune-in loyalty in various DAYPARTS. The phrase is often used synonymously with STRIPPING.

act   a section of a television program that is part of the organization and construction of the show. Half-hour SITCOMS are often divided into three segments, with the first act establishing the problem, the second complicating it, and the third resolving it.

Action for Children's Television (ACT)   a Boston-based nonprofit organization that promoted and advanced standards and excellence in children's programming. The national citizens' group was instrumental in forcing the networks, stations, the FEDERAL COMMUNICATIONS COMMISSION (FCC), the FEDERAL TRADE COMMISSION (FTC), and the industry at large to curb excesses in children's programming and in the COMMERCIALS directed at them. It disbanded in 1992.

A/D conversion   a process that converts an ANALOG COMMUNICATIONS signal into its equivalent in DIGITAL COMMUNICATIONS terms. A converter measures the AMPLITUDE (voltage) of an incoming analog electronic signal thousands of times each second and stores each of the measurements as a number.

addressable converter   a small box containing a microcomputer that is placed on the television set. The subscriber uses a key pad connected to the converter and by punching buttons, orders a particular channel from the cable HEADEND. Such converters have been labeled "addressable" because they can be individually accessed by the computer at the headend.

adjacency a period of COMMERCIAL TIME that precedes or follows a network program. About two minutes in length, it is offered to advertisers for SPOT ANNOUNCEMENTS.

Adult Learning Service (ALS) a department of the PUBLIC BROADCASTING SERVICE (PBS). This service acquires and transmits TELECONFERENCES and TELECOURSES.

adult videos a euphemistic term used to identify films in the prerecorded home video industry whose content is primarily sexual. The films are divided into "softcore" and "hardcore" titles, with the latter being more sexually explicit.

advanced television (ATV) a broad term that refers to the technical systems that will eventually improve television image and sound. ATV encompasses a number of different technical proposals that are known by the initials used by their proponents. They include ACTV (advanced compatible TV), ADTV (advanced definition TV), IDTV (improved definition TV), VHDTV (very high definition TV), EDTV (enhanced definition TV), and HDTV (high definition TV). The most widely

used are the EDTV and HDTV nomenclature, and the initials HDTV are often used interchangeably with ATV.

Advanced Television Evaluation Laboratory (ATEL) a Canadian organization that was established to assist the CANADIAN DEPARTMENT OF COMMUNICATIONS (DOC) and the Canadian television industry in selecting a new ADVANCED TELEVISION (ATV) system in conjunction with other tests conducted by the ADVANCED TELEVISION TEST CENTER and CABLELABS in the United States. The activities of these three test organizations were coordinated by the ADVISORY COMMITTEE ON ADVANCED TELEVISION SERVICE (ACATS) of the FEDERAL COMMUNICATIONS COMMISSION (FCC) in the United States and by the Advanced Broadcast System of Canada, and advisory group to the Canadian DOC.

Advanced Television Systems Committee (ATSC) a private volunteer committee of manufacturers of video equipment and receivers. ATSC was formed by electronics companies in the United States and the SOCIETY OF MOTION PICTURE AND TELEVISION ENGINEERS (SMPTE) in response to proposals for production standards for an ADVANCED TELEVISION system (ATV) by the Japan Broadcasting Corporation (NHK).

Advanced Television Test Center Inc. a private, nonprofit corporation organized by the television broadcasting industry to examine the options for a new U.S. terrestrial transmission standard for an AD-VANCED (ATV) TELEVISION service. The center's primary focus is on national efforts to establish a single terrestrial ATV transmission standard. (See also DIGITAL VIDEO COMPRESSION, ENHANCED NTSC SYSTEMS, and SIMULCAST HIGH DEFINITION TELEVISION.)

Advertiser Syndicated Television Association (ASTA) a New York-based trade association that promotes the growth of advertiser-supported SYNDICATION.

advertising All forms of commercial and noncommercial messages designed to persuade or inform an audience. Advertising is usually developed to promote the purchase of goods or services but it can also be used to influence thoughts and ideas. In cable and television, it takes the form of COMMERCIALS and SPOTS (COMMERCIAL TIME).

advertising agency an independent company that creates and manages advertising for other firms. In addition to creating and producing COMMERCIALS, ad agencies conduct product and consumer research and buy COMMERCIAL TIME for CLIENTS on television and cable operations and on radio and in newspapers and magazines. (See also ACCOUNT, ADVERTISING AGENCY COMMISSION, BOUTIQUE AGENCY, COMMISSION-ABLE, and FULL-SERVICE ADVERTISING AGENCY.)

advertising agency commission the traditional means of compensat-

ing ad agencies. The conventional compensation arrangement has been for the advertiser to pay the agency the full amount charged for COMMERCIAL TIME or space by the MEDIA. The television or cable operation, however, gives the ad agency a discount and the agency keeps the difference between the full amount and the discount as its compensation. The traditional discount has been 15 percent.

advertising agency review the procedure followed when an ACCOUNT is opened to competition from other advertising agencies. When a contract between an agency and a CLIENT expires, it is often renewed for another period. On occasion, however, the client believes there is need for a change and undertakes an agency review. While the incumbent is usually also encouraged to make a PRESENTATION, other agencies are contacted to PITCH the account.

Advertising Council Inc. (ACI) a nonpolitical, nonpartisan, volunteer organization dedicated to public interest causes. It is composed of prominent advertising executives who donate their creative talents and time to produce and nationally distribute PUBLIC SERVICE ANNOUNCEMENTS (PSA).

Advertising Federation of America (AFA) an association of advertising organizations. AFA is dedicated to networking and to the exchange of ideas and information.

Advertising Research Foundation (ARF) an organization founded in

1936 by the AMERICAN ASSOCIATION OF ADVERTISING AGENCIES (4AS) and the ASSOCIATION OF NATIONAL ADVERTISERS (ANA). The group includes those organizations, plus ADVERTISING AGENCIES, universities, and media companies.

advertorial a combination of the words "advertisement" and "editorial." The term refers to a COMMERCIAL on a cable or television operation advocating a particular point of view. The advertiser pays for the production and the COMMERCIAL TIME.

An advertorial differs from a PUBLIC SERVICE ANNOUNCEMENT (PSA) inasmuch as a PSA does not normally advocate a particular or controversial position, and the transmission time is donated by the cable or television operation.

Advisory Committee on Advanced Television Service (ACATS) a committee established by the FEDERAL COMMUNICATIONS COMMISSION (FCC) in 1987. It was given a mandate to study and recommend a new ADVANCED TELEVISION (ATV) system for terrestrial television services in North America. It consisted of 25 leaders in broadcast, cable operations, program production, and television-set manufacturing. They were assisted by many volunteers and three testing centers: the ADVANCED TELEVISION TEST CENTER INC., CABLELABS, and the ADVANCED TELEVISION EVALUATION LABORATORY.

affiliate a type of video retail store connected to a larger, national entity that uses its name and image. The headquarters company provides advertising and marketing support and other services similar to those provided by firms that offer FRANCHISES. Affiliate operations, however, are not regulated by the FEDERAL TRADE COMMISSION (FTC). They charge no monthly fees; their income is derived from the MARGINS they receive in buying and distributing products from WHOLESALERS.

affiliated station a type of broadcast station in a local community that is an outlet for a major NETWORK. The station signs an AFFILIATION CONTRACT with the network to air programs originated by that organization. Affiliated stations usually receive more than 70 percent of their programming from their network. The networks provide the programs at no cost to their affiliates in exchange for COMMERCIAL TIME. (See also INDEPENDENT STATIONS, O & O, and PREEMPTION.)

affiliation contract the formal statement of the relationship between an AFFILIATED STATION and its network. The station is compensated for the carriage of the programs according to the terms of its affiliation contract. It details the relationship, which is partially regulated by FEDERAL COMMUNICATIONS COMMISSION (FCC) rules.

Ag*Sat a satellite-based, nonprofit consortium. It is based at the Nebraska PUBLIC TELEVISION (PTV) STATE NETWORK headquarters in Lincoln, and it transmits TELECONFER-

ENCES and live agriculture-related credit TELECOURSES to 33 land-grant colleges in 13 states.

AGB Television Research a ratings company that had the exclusive rights to determine the number of television viewers in the United Kingdom from 1968 to 1990. It now shares those rights with another company but also provides audience research services in 15 other countries. The company invented the PEOPLE METER.

Agency for Instructional Technology (AIT) a nonprofit organization dedicated to the improvement of instruction through the use of electronic media. The agency is made up of 60 state and provincial educational organizations from the United States and Canada that have banded together to develop and create high quality INSTRUCTIONAL TELEVISION (ITV) programming, based on careful research and curriculum needs.

air check an OFF-THE-AIR videotaped recording of a COMMERCIAL or a program as it is actually seen on a television station, network or cable system. It is used by a SPONSOR to place the piece in context and to evaluate its talent, writing, production values and effectiveness. Air checks are also made at the direction of an ADVERTISING AGENCY to verify that a commercial has been aired.

air date the date (day) on which a television program or COMMERCIAL is scheduled to be transmitted (or "aired"). The term originated in ra-

dio and has been carried over into television broadcasting and cable operations.

all-channel law a 1962 amendment to Section 303 of the COMMUNICATIONS ACT OF 1934. This law required television set manufacturers to equip all receivers with UHF as well as VHF tuners by 1964. A viewer could then theoretically easily receive any local television broadcast signal. Today, all receivers have built-in UHF and VHF tuners, but with the advent of cable television, the distinction between UHF and VHF stations has become less important because they all appear equal on cable channels.

Alliance of Canadian Cinema, Television, and Radio Artists (ACTRA) a Canadian trade union representing actors, performers, announcers, cartoonists, singers, puppeteers, writers, extras, hosts, and models in motion pictures, radio and television programs, and in commercials throughout Canada. This single union combines most of the activities and representation of four U.S. unions.

Alliance of Motion Picture and Television Producers a Hollywood-based organization that represents the major studios and independent production companies in collective bargaining with the trade unions and societies.

allocations the blocks of FREQUENCIES apportioned to specific communications services to avoid interference with other transmissions. The electromagnetic spectrum is divided into groups of frequencies, and these are then allocated to specific communication services, such as AM and FM radio and UHF and VHF television.

Within the United States the FEDERAL COMMUNICATIONS COMMISSION (FCC) allots frequencies and channels to specific geographic locations. Because cable transmissions do not use the broadcasting spectrum, no such specific allocation system exists in that industry. (See also SIXTH REPORT AND ORDER.)

Alpha Chi Alpha   See SOCIETY FOR COLLEGIATE JOURNALISTS.

Alpha Epsilon Rho (AERho) an honorary college fraternity, founded in 1943, that holds seminars and meetings to further the members' understanding of the MEDIA. The organization is composed of students and faculty in broadcasting, cable and related fields.

Alpha Phi Gamma   See SOCIETY FOR COLLEGIATE JOURNALISTS.

alternate sponsorship   the practice of two or more companies alternating as SPONSORS on a day-by-day basis. This technique has been traditionally used for program series and SOAP OPERAS or for programs in SYNDICATION that are transmitted on a regular basis during the week. (See also ACROSS THE BOARD and CHECKERBOARDING.)

alternate weeks sponsorship an advertising practice in which a COM-MERCIAL on a television or cable operation is scheduled for one week, then dropped for a week, and resumed the third week. This technique is sometimes used in small or limited CAMPAIGNS.

alternative   dispute   resolution (ADR) a method of settling conflicts through negotiation that is used by the FEDERAL COMMUNICATIONS COMMISSION (FCC) and other federal agencies. ADR includes the gathering of parties for negotiated rulemaking, neutral evaluation and fact-finding, binding and nonbinding arbitration, and several types of mock trials.

alternative television a type of television that, in its earliest manifestation, was called "underground" or "guerrilla TV." It referred to the activist, radical movement in television production, which often had a political purpose. Collective groups such as Global Village and Videofreex produced antiestablishment programs that usually explored left-wing political ideas. Many of the counterculture programs were aligned with the art world. As cable emerged, alternative video found outlets on public access and later CUPU LEASED ACCESS CHANNELS and PEG CHANNELS.

American Advertising Federation (AAF) a nonprofit association of people in advertising. The AAF promotes standards for responsible advertising in all media and encourages education in the field.

American Advertising Museum an institution concerned with artifacts and documents related to the role of advertising in American culture. The museum's public collection includes more than 200,000 print advertisements from the 1890s onward, 44,000 slides of billboards, and an extensive library. It is located in Portland, Oregon.

American Association of Advertising Agencies (4As or AAAA) a prestigious nonprofit association, composed of the major ADVERTISING AGENCIES in the United States and commonly known as the 4As. Membership is by election only, and its now 755 member agencies reportedly handle 80 percent of all national advertising.

American Association of Media Specialists and Librarians (AAMSL) an association whose members are involved in AUDIOVISUAL COMMUNICATIONS. The organization evaluates media resources and educational technologies in order to help improve instruction in the classroom.

American Broadcasting Company (ABC) See ABC TELEVISION NETWORK and CAPITAL CITIES/ABC INC.

American Council for Better Broadcasts See NATIONAL TELEMEDIA COUNCIL.

American Electronics Association (AEA) a nonprofit association composed of companies in the electronic components and equipment business. The AEA membership is engaged in the manufacture, research, and development of technology.

American Family Association a nonprofit watchdog organization with two major purposes. It seeks to persuade the networks to eliminate sex, profanity, violence, and an anti-Christian bias from their programming, and to encourage the enforcement of state obscenity laws related to hardcore and softcore pornography. The group was founded in 1977 as the National Federation for TV Decency. (See also CLEAR-TV.)

American Federation of Musicians of the U.S. and Canada (AFM) a labor union for professional musicians that is affiliated with the AFL-CIO and represents instrumental musicians in appearances, including performances on television.

American Federation of Television and Radio Artists (AFTRA) a union that represents newscasters and sportscasters, singers and dancers, announcers, actors, and other performers. AFTRA contracts cover network and local New York television appearances in commercials and dramatic and nondramatic programs and in industrial and nonbroadcast prerecorded programs throughout the United States.

American Film and Video Association (AFVA) a nonprofit association that is dedicated to furthering the development of NONTHEATRICAL FILM and video programming. The membership consists of library and AUDIOVISUAL COMMUNICATIONS centers and producers and distributors of DOCUMENTARIES and educational and informational film and video programs.

American Film Institute (AFI) a Washington D.C.-based nonprofit organization whose purpose is to help develop video and film programs in the United States. AFI makes grants to small independent producers, awards internships, sponsors competitions, and organizes video and film festivals. (See also NATIONAL CENTER FOR FILM AND VIDEO PRESERVATION.)

American Film Marketing Association (AFMA) a trade association that sponsors the annual spring American Film Market, where producers of independent motion pictures exhibit their new product to theatrical and home video distributors.

American Museum of the Moving Image (AMMI) a nonprofit museum devoted to the art, technology, history and social impact of the moving image. The 50,000-square-foot complex, located in New York City, features displays of some 60,000 objects, including early movie cameras and television sets. It differs from the MUSEUM OF TELEVISION AND RADIO and the MUSEUM OF BROADCASTING COMMUNICATIONS (MBC) in its inclusion and underlining of theatrical film and film memorabilia. (See also ATAS/UCLA TELEVISION ARCHIVES, LIBRARY OF CONGRESS: MOTION PICTURE, BROADCAST, RECORDED SOUND

DIVISION, and NATIONAL CENTER FOR FILM AND VIDEO PRESERVATION.)

American Research Bureau, Inc. (ARB) an audience research company that was the primary source in providing local television stations with program RATINGS from 1949 to 1982. The reports, called "ARBs," were later expanded to cover all markets. (See also ARBITRON COMPANY.)

American Society for Healthcare, Education, and Training (ASHET) a Chicago-based affiliate of the American Hospital Association (AHA) that is dedicated to the improvement of education and training in health-care facilities. Much of the emphasis is on the use of AUDIOVISUAL COMMUNICATIONS technology.

American Society for Training and Development (ASTD) the premier membership association for professionals in education and training in business and industry. This organization includes many members who utilize AUDIOVISUAL COMMUNICATIONS and instructional technology in training and development for employees within a company.

American Society of Composers, Authors, and Publishers (ASCAP) a nonprofit, performance-rights association of music publishers, composers, and lyricists, which acts as a licensing agency for much of the music performed on television, cable, and video. It licenses the performance rights of copyrighted music (live or recorded) but it does not license the mechanical rights for recording nor the rights to reproduce the sheet music; those rights are often handled by the Harry Fox Agency or SESAC INC. (See also BROADCAST MUSIC INC. [BMI].)

American Society of Educators See AMERICAN ASSOCIATION OF MEDIA SPECIALISTS AND LIBRARIANS (AAMSL).

American Society of TV Cameramen (ASTVC) a national, nonprofit organization that is composed of camera operators and former camera operators. The society strives to foster communication among camera operators in the industry and operates the International Society of Videographers (ISV).

American Sportscasters Association (ASA) a nonprofit association composed of radio and television sportscasters. The group administers the Sportscasters' Hall of Fame.

American Telephone and Telegraph (AT&T) a giant private company that has played an important role in radio and television in the United States. A business created by Alexander Graham Bell's invention of the telephone in 1876, AT&T became a monolith monopoly in long-distance communications, creating a transcontinental telephone service by 1915. "Ma Bell" manufactured and rented all telephones in the United States (through a subsidiary) and also operated the nation's long-distance service and all local telephone companies.

AT&T entered the radio broadcasting business during the early 1920s, manufacturing transmitters and owning and operating some radio stations. The early radio networks CBS and NBC rented AT&T lines. AT&T also provided the three commercial television networks with their terrestrial interconnection system until the system was largely replaced by SATELLITES in the 1970s. The company developed the EARLY BIRD SATELLITE (launched by COMMUNICATIONS SATELLITE CORPORATION [COMSAT]) and succeeding generations of the Telstar series.

The U.S. Justice Department eventually filed an antitrust suit against the company, and in 1982 AT&T and the Department finally settled out of court. The national Bell system was dismantled and its local services reorganized into seven independent regional firms.

AT&T still supplies some terrestrial national interconnections for the television networks, but the company's current emphasis is on long-distance telephone service, information exchange via computers, and cable and satellite communications.

American Television and Communications Corporation (ATC) a cable communications company that is one of the nation's top 10 MULTIPLE SYSTEM OPERATORS (MSO). Founded in 1967, it brings cable services to subscribers in 33 states. More than 60 of its systems have local production facilities.

The majority of the company's stock was previously owned by Time

Inc. After that company's acquisition of Warner Communications Inc. in 1989, ATC became a subsidiary of the new TIME WARNER INC. ATC and Warner Cable Communication are now a part of the Time Warner Cable Group, which was formed to manage the newly merged company's cable operations.

American Women in Radio and Television (AWRT) a nonprofit membership association that encourages the growth and development of women in the radio and television industries. With 52 local chapters worldwide, it hosts an annual convention and sponsors a number of awards.

America's Public Television Stations (APTS) a lobbying organization that represents the interests of PUBLIC TELEVISION (PTV) stations before the CORPORATION FOR PUBLIC BROADCASTING (CPB), federal agencies, Congress, and the general public, and provides planning and research services for its member stations. The organization is a part of the ASSOCIATION FOR PUBLIC BROADCASTING (APB).

amplifier a device used to make electronic signals stronger. In the process of amplification, signal strength is increased to provide greater power for subsequent use. Some amplifiers increase the voltage or current, while others amplify the power of the signal. (See ATTENUATION, BRIDGING AMPLIFIER, DISTRIBUTION AMPLIFIER, and PREAMPLIFIER.)

amplitude a description of the level of an electronic signal. Amplitude is the strength or magnitude of a waveform or voltage. In audio, this intensity governs the loudness of the sound; in video, the brightness of the image. (See also AMPLITUDE MODULATION.)

amplitude modulation (AM) a process by which the AMPLITUDE of an otherwise constant signal is altered. In the United States, amplitude modulation is used for broadcast radio transmission (AM radio) and for the video portion of a television signal. The amplitude modulation produces a signal that occupies a little more than twice the bandwidth of the modulating signal. (See also ANALOG COMMUNICATIONS, ANTENNA, FREQUENCY MODULATION (FM), and TRANSMITTER.)

analog communications a means of communicating electronically. Sound waves are converted to electrical signals to be stored, amplified, and transmitted. A continuously varying voltage is used to represent the characteristics (at any instant) of the sound waves being processed.

The magnitude of the signal at any given instant is analogous to the magnitude of the sound wave that is being communicated. The same applies to video signals where the magnitude of the signal represents the brightness of the pixel being transmitted.

Analog transmission and reception schemes have been the norm in broadcasting since the inception of the concept, but these systems are rapidly being replaced by DIGITAL COMMUNICATIONS, which are not subject to the distortion and interference that plague the older method. (See also A/D CONVERSION.)

Anik satellites   a series of Canadian domestic satellites used as a transcontinental interconnection system. The satellites also provide television and telephone services to the sparsely populated northern part of Canada.

animatic   a rough draft of a television COMMERCIAL, which is developed by taking photographs of individual STORYBOARD sketches and assembling them into a film strip. The audio portion is recorded on tape and synchronized to the visual images, resulting in an animatic.

animation   an appearance of movement that is created from still drawings or objects. Hundreds of drawings or movements of the objects are required and each still frame must be shot individually. The technique can be accomplished either by a special film camera or by computers. The human eye makes the static item or graphic appear to come alive with movement.

Annenberg/CPB projects   important INSTRUCTIONAL TELEVISION (ITV) projects that were funded by the Annenberg Foundation in 1981 and administered by the CORPORATION FOR PUBLIC BROADCASTING (CPB). TELECOURSES were produced for college students and adults and were aired as a part of the ADULT LEARNING SER-

VICE (ALS) of the PUBLIC BROADCASTING SERVICE (PBS). (See also DISTANCE EDUCATION, and OPEN UNIVERSITY.)

answer print   a second-to-last stage in the technical production of a filmed program or commercial in which the audio and video portions are combined with all of the optical effects (such as SUPERIMPOSITIONS and DISSOLVES) into a nearly finished version. The term is derived from the process of sending a WORKPRINT to the film laboratory and receiving an answer print back, which is then reviewed and approved before a FINAL PRINT is ordered.

antenna   a device used for transmitting or receiving electromagnetic signals. In television broadcasting, a transmitting antenna is a metallic device mounted on a tower that is located on high ground or on the top of a tall building. Receiving antennas are the traditional rabbit ears or dipole antennas on the top of a television set or rooftop, or simple monopole (single) metal units 12 inches in height.

A receiving antenna is also an integral part of a TELEVISION RECEIVE ONLY (TVRO) dish. The large parabolic unit captures SATELLITE signals and focuses them onto a small antenna that is located at the center of the dish. (See also ANTENNA HEIGHT ABOVE AVERAGE TERRAIN and EFFECTIVE RADIATED POWER [ERP].)

antenna height above average terrain   one way of measuring TV antenna height. The extent of a television broadcast signal depends on the FREQUENCY at which the signal is transmitted, the type and EFFECTIVE RADIATED POWER (ERP) of the TRANSMITTER, and the height of the ANTENNA.

Antenna height is measured in two ways: height above ground and height above average terrain. The height of an antenna above the average terrain is authorized by the FEDERAL COMMUNICATIONS COMMISSION (FCC) as a part of its permission to broadcast, and the height above ground is authorized by the FCC and the Federal Aviation Authority (FAA) to ensure that the tower and antenna do not constitute a hazard for aircraft.

anthology programming   self-contained programs that begin and end during the period of time they are on the air and do not carry over to subsequent shows. Although the anthology technique is used in all subject areas from music to comedy, it is most commonly applied to drama programs.

Antiope a one-way broadcast French TELETEXT system that carries news, weather, and stock market reports, along with other information. It was developed as a part of a national program called TELEMATIQUE and has a two-way cable VIDEOTEXT companion.

Arbitron Company a media research company that is now best known for its radio audience research. It provides audience measurement data to broadcasters and ADVERTISING AGENCIES. The company's initial RATINGS were known

as ARBs, after the original company name, the AMERICAN RESEARCH BUREAU INC. The company started in television in 1949 and added radio measurement in 1965. For 44 years it was the major competitor of A. C. NIELSEN in providing television ratings. Arbitron ratings were based on 200 marketing areas, each called an AREA OF DOMINANT INFLUENCE (ADI). These geopolitical entities were similar to Nielsen's DESIGNATED MARKET AREA (DMA) classifications. The industry tended to rely on Arbitron for the measure of the local stations' audiences and Nielsen ratings for network programs. In 1994, however, Arbitron discontinued its television rating service.

arc a camera movement in which the mount and camera are moved in a curving path to the right or left of the subject. The movement is a combination of a DOLLY IN and a TRUCK shot. The term is also used as shorthand for an arc light or carbon-arc light, the high-intensity lighting instrument used to provide light over vast areas and in large auditoriums and arenas.

area of dominant influence (ADI) the MARKET area comprised of the counties surrounding a metropolitan center (as defined by the ARBITRON COMPANY). The major viewing audience for the television stations in that particular area was within the ADI. About 200 such geopolitical areas were so designated in the United States.

ADIs were similar to the DESIGNATED MARKET AREAS (DMAs) of the A. C. NIELSEN COMPANY. (See also

CONSOLIDATED METROPOLITAN STA-
TISTICAL AREA [CMSA] and METROPOL-
ITAN STATISTICAL AREA [MSA].)

Armed Forces Broadcasters Association (AFBA) a nonprofit association that provides an opportunity for camaraderie among its membership of former as well as current military broadcasters throughout the world.

Armed Forces Radio and Television Service (AFRTS) a service that operates more than 800 radio and television outlets in 56 countries. The stations are established at locations with concentrations of U.S. forces overseas, in U.S. territories, and aboard U.S. Navy ships.

Army–McCarthy hearings televised congressional hearings in 1954 that hastened the end of MCCAR-THYISM. The senator from Wisconsin, Joseph McCarthy, had been conducting his flamboyant witchhunts for communist sympathizers in all areas of government. He finally focused on the U.S. Army. Senate committee hearings on his charges of subversion in that branch of the military were held and televised beginning in April 1954. During the hearings, McCarthy revealed himself as a ranting demagogue. As a result of his posturing and tirades, the Senate voted later that year to censure him.

Arts and Entertainment Network (A&E) a basic cable network designed to bring quality programming in comedy, drama, documentary, and the performing arts to discrimi-nating viewers. The viewership is decidedly upscale, professional adults who are usually active in their communities. The network has won more ACE (Award of Cable Excellence) AWARDS than any other basic cable network. It is owned by the HEARST CORPORATION, CAPITAL CITIES/ABC, and NATIONAL BROADCASTING CORPORATION.

Asian Broadcasting Union (ABU) a nonprofit organization that encourages the use of media to aid international understanding. The members of the ABU are national broadcasting organizations in the region.

aspect ratio the ratio between the width and height of the television picture. The proportions of the picture and pickup tube in current television sets and television cameras is 1 to 1.33, or 3:4. The picture, therefore, is one and one-third times as wide as it is high. In contrast, most theatrical motion picture screens are nearly twice as wide as they are high, with an aspect ratio of about 9:16.

Aspen Institute Program on Communications and Society a nonprofit project dedicated to enhancing the social benefits of new technology. In the past, the program has sponsored and conducted research on major issues in the communications area, specifically recommending courses of action and policy in the field of television.

Aspen Institute Rulings actions by the FEDERAL COMMUNICATIONS COMMISSION (FCC) that opened the way for the regular telecasting of two-person debates between political candidates. Concerned with the requirements of the EQUAL TIME (OPPORTUNITY) RULES, the Commission had previously held that every single candidate for a political office had to be accommodated in a debate format. A televised debate could be crowded with as many as seven or eight people vying for the same office, most of them from minor political parties.

At the instigation of the ASPEN INSTITUTE PROGRAM ON COMMUNICATIONS AND SOCIETY, the Commission overruled its stipulation regarding debates in 1975. Under the new rules, the FCC held that under certain conditions, the broadcasts of two-person debates between political candidates could be considered exempt from the "equal opportunity" provisions.

Association for Communication Administration (ACA)   a nonprofit organization that consists of deans, directors, and the chairs of schools, divisions, or departments of communication in the nation's colleges and universities. The ACA is affiliated with the SPEECH COMMUNICATION ASSOCIATION (SCA).

Association for Education in Journalism and Mass Communication (AEJMC)   the oldest nonprofit association in journalism education in the United States. The goal of this professional organization of college and university journalism professors is to improve the techniques of teaching and to encourage research in the field.

Association for Educational and Training Technology (AETT)   a nonprofit membership organization in the United Kingdom that is devoted to the improvement of education and training through AUDIOVISUAL COMMUNICATIONS. It is similar to the ASSOCIATION FOR EDUCATIONAL COMMUNICATIONS AND TECHNOLOGY (AECT) in the United States and the ASSOCIATION FOR MEDIA AND TECHNOLOGY IN EDUCATION IN CANADA (AMTEC).

Association for Educational Communications and Technology (AECT)   a national association of people involved in AUDIOVISUAL COMMUNICATIONS that is concerned with learning and educational technology and acts as a clearinghouse and information center about educational and instructional media. AECT members include media and library specialists, university researchers and professors, government media personnel, and school specialists in educational media.

Association for Information Media and Equipment (AIME)   a trade association whose membership consists of companies and organizations engaged in the production or distribution of NONTHEATRICAL FILM/video programs, as well as the manufacturers of equipment used in AUDIOVISUAL COMMUNICATIONS operations.

Association for Maximum Service Television (MSTV)   a nonprofit association whose mission is to support free, high-quality, and community-oriented television service throughout the United States. It also promotes government action to maintain high quality TV transmission, and informs the industry and the general public about changes that could affect the technical quality of television.

Association for Media and Technology in Education in Canada (AMTEC)   a nonprofit membership organization similar to the ASSOCIATION FOR EDUCATIONAL COMMUNICATIONS AND TECHNOLOGY (AECT) in the United States and the ASSOCIATION FOR EDUCATIONAL AND TRAINING TECHNOLOGY (AETT) in the United Kingdom. Its objectives are to identify critical issues and developments in educational media and technology and to provide the dissemination of valid information.

Association for Public Broadcasting (APB)   a nonprofit organization that is the parent of AMERICA'S PUBLIC TELEVISION STATIONS (APTS). The tax-exempt organization collects grants and contributions and conducts research and planning for the PUBLIC TELEVISION (PTV) stations in the United States.

Association of Audio-Visual Technicians (AAVT)   a nonprofit professional organization of people involved in the operation and maintenance of audiovisual equipment such as videocassette machines and slide, overhead, filmstrip, and 16mm film projectors.

Association of Black Motion Picture and Television Producers   a nonprofit organization that consists of qualified black theatrical film and TV producers.

Association of Canadian Television and Radio Artists   See ALLIANCE OF CANADIAN CINEMA, TELEVISION, AND RADIO ARTISTS (ACTRA).

Association of Catholic TV and Radio Syndicators   an organization that allows its members to exchange information and discuss SYNDICATION from a Roman Catholic viewpoint. The association is affiliated with UNDA-USA.

Association of Cinematographers, TV, and Allied Technicians (ACTAT)   a British trade union that represents production personnel in television and film production in the United Kingdom. The union represents members working in feature films, television programs, documentaries, specialized productions, and animated shows.

Association of FCC Consulting Engineers   an organization consisting of registered consulting engineers who practice before the FEDERAL COMMUNICATIONS COMMISSION (FCC). The members exchange information and ideas about engineering and channel allocation issues.

Association of Independent Commercial Editors (AICE)   an organization that acts as a venue for dealing with the common interests and problems of the companies and people that edit film and videotape COMMERCIALS.

Association of Independent Commercial Producers (AICP) a trade association that serves as a bridge between the producers of COMMERCIALS and ADVERTISING AGENCIES and their CLIENTS. The organization also represents companies involved in commercial production with local, state, and national governmental agencies and with unions within the industry.

Association of Independent Television Stations (INTV) a nonprofit organization whose membership consists of INDEPENDENT STATIONS that are not affiliated with the three major networks. Most of the members operate UHF stations and some are GROUP BROADCASTERS. The association represents its members before Congress, the FEDERAL COMMUNICATIONS COMMISSION (FCC), and the public.

Association of Independent Video and Film Makers (AIVF) a trade association that represents the interests of the smaller independent video and film makers before the public and within the industry. The AIVF sponsors professional meetings, screenings, and seminars; encourages the distribution of its members' work; and monitors the status of independent productions on PUBLIC TELEVISION (PTV), PEG CHANNELS, and CUPU LEASED ACCESS CHANNELS.

Association of Maximum Service Telecasters (AMST) See ASSOCIATION FOR MAXIMUM SERVICE TELEVISION (MSTV).

Association of National Advertisers (ANA) a trade association that represents national advertisers in all MEDIA, including television, cable, and video. The membership consists of companies that advertise nationally, and the organization monitors advertising practices in the United States in the interests of its members.

Association of Visual Communicators (AVC) a nonprofit membership organization that encourages and promotes excellence within the AUDIOVISUAL COMMUNICATIONS, film, and video industries. The membership consists of technical and creative people and managers in the industry.

asynchronous a technical term that describes signals that are not synchronized to one another, such as the sound with the picture or two video signals of the same scanning standard that are not in SYNC with each other.

ATAS/UCLA Television Archives a program archive operated by the University of California at Los Angeles (UCLA) in conjunction with the ACADEMY OF TELEVISION ARTS AND SCIENCES (ATAS). It contains one of the largest collections of programs on KINESCOPE, film, and videotape in the United States.

ATS-6 a 1974 communications SATELLITE used to experiment with the educational uses of the new technology. Demonstration projects using the new device were designed to bring quality INSTRUCTIONAL TELEVI-

SION (ITV) programming to rural communities.

attenuation the loss in electrical power between a signal's original transmission and its reception. The loss can result from the length of the transmission through the air or via wire or COAXIAL CABLE. Attenuation may also be caused by the quality or type of equipment used in transmitting or receiving the signal. The loss of power is usually expressed in DECIBELS (DB).

auction a televised fundraising event in the PUBLIC TELEVISION (PTV) industry. Different types of merchandise or services are donated by individuals or commercial firms and are placed in the studio. Viewers are invited to bid on specific items by phoning the station, and the highest bidder in a given time period gets the merchandise.

audience flow the movement of television viewers over a period of time. Actions such as turning on the set, changing channels, and turning off the set determine the audience flow. Viewers' traffic behavior among programs is extremely important to many advertisers. It is measured by RATING companies such as A. C. NIELSEN.

Audimeter an audio audience-measuring device that was adapted to measure television viewing in 1950. In 1971 an improved version called an SIA (Storage Instantaneous Audimeter) allowed data to be collected and transmitted electronically.

The Audimeter is now used as the minicomputer brain to store data collected by a PEOPLE METER.

audio sound that can be heard by the human ear, including noise, music, and speech. It is the sound portion of television, and is usually considered to be in the FREQUENCY range of normal human hearing, which is between 15 HERTZ (HZ) and 20 KILOHERTZ.

Audio Engineering Society (AES) a membership organization that serves the interests of engineers and technicians who operate recording equipment in radio, television, and motion pictures and in music recording studios.

Audio-Visual Management Association (AVMA) an elite, by-invitation-only, nonprofit organization composed of the managers of the AUDIOVISUAL COMMUNICATIONS departments of major companies.

audiographics an interactive DISTANCE EDUCATION tool that uses computers to transmit graphics and audio simultaneously over common telephone lines. Teachers and students at different locations can observe and share still images, slides, and comments before, during, or after a lesson. While the system does not permit the transmission of full motion pictures, it does offer a relatively inexpensive method of teacher-pupil interaction.

audiovisual communications an almost obsolete term that is used in the field of education and training to describe the development, production, transmission, and use of sight and sound as a contribution to learning.

During World War II millions of civilian recruits were trained and retrained for wartime duties, using audiovisuals. These included globes, models, mockups, charts, maps, magnetic boards, overhead and opaque projectors, posters, slides, and 16mm film.

After the war, the term "audiovisual instruction" described the more systematic use of the materials in a formal school environment. Later, EDUCATIONAL TELEVISION (ETV) became a part of the mixture and the terms MEDIA and "educational media" became popular, to accommodate its role.

The proliferation of phrases meaning roughly the same thing created considerable confusion in the educational establishment. Some began to use the phrase "instructional technology," but many believed the term placed too much emphasis on the hardware or the devices that were used, and not enough on the process. Those who favored the process preferred to use some variation of the term "media."

In addition to the traditional AV tools, teachers now use AUDIOGRAPH-ICS, CD–ROM, INSTRUCTIONAL TELE-VISION (ITV) SATELLITES (STAR SCHOOLS PROGRAMS), INSTRUC-TIONAL TELEVISION FIXED SERVICE (ITFS), INTERACTIVE VIDEO, and MUL-TIMEDIA combinations in the classroom. Today, the term "audiovisual communications" has become old-fashioned and is seldom used to describe the concept and the field. The term "educational media" has gained favor among professionals as a broad, all-encompassing phrase with "instructional media" often applied in a more systematic, formal, teaching situation.

audition a critical hearing that is scheduled before casting a television show or film production. A performer tries out for a part by demonstrating talent or suitability in a brief trial performance.

The term is also used to describe a distinct, separate, audio circuit that allows an engineer to preview or audition sound before using it in a production. Music from a tape or record can be cued up or readied for insertion into a program in this manner.

automated videocassette systems videocassette systems that are computer-controlled to play back videotapes of commercials and programs all day and night, in the proper sequence, without error. Sometimes called "library management systems," the automated start/stop and selection process is controlled by computers.

automatic fine tuning (AFT) an electronic circuit that automatically tracks a particular FREQUENCY. A common part of videocassette recorders and other consumer electronic equipment, it corrects slight drifts and maintains signal strength at a maximum level.

automatic gain control (AGC) an electronic circuit that automatically controls the operating level in an AM-PLIFIER. It monitors the output of an amplifier, ensuring that the outgoing signal is consistent, regardless of any variations in the level of the input signal.

automatic number identification (ANI) a telecommunication process often used in PAY-PER-VIEW (PPV) circumstances, ANI allows the automatic identification of the originating telephone (by number) at the reception end of a call.

availabilities the COMMERCIAL TIME periods that have not yet been sold by television or cable operations. An advertiser or ADVERTISING AGENCY will query the local station or the national STATION REPRESENTATIVE (REPS), to determine which time slots are available for purchase from the INVENTORY of the operation.

average audience (AA) a measurement of the television and cable audience. Familiarly known as the AA rating, it is an estimate of a program's average audience on a minute-by-minute basis. A part of the measurement in the NIELSEN TELEVISION INDEX (NTI), this method is also used in developing the audience SHARE and the COST PER THOUSAND (CPM) for advertisers and agencies.

B

"B" title a motion picture within the home video industry that features less-than-name actors but has had a reasonably successful theatrical run. The designation is applied to indicate the anticipated retail demand for the film and is distinct from the term "B movie," which is a measure of relative quality.

back-to-back scheduling a type of television programming strategy that involves the scheduling of similar shows consecutively. Two programs appealing to the same DEMOGRAPHICS are aired successively to encourage the audience to stay tuned.

background lighting lights that illuminate the walls of a room or set and generally separate them from the subject in a television or film production. These types of lights provide flat lighting, usually controlled by a DIMMER, that enhances the visibility of a CYCLORAMA or background. SCOOP LIGHTS and SOFT LIGHTS are used to provide the overall light.

background noise signal-path voltages that are not related to the signal. These usually random fluctuations in voltage mix with the desired signal and are amplified with it, emerging as "hiss" in audio circuits and "snow" in video.

backlight a television and film lighting technique that directs light onto a performer, object, or scene from behind rather than from the front. It is used to separate the subject from the background.

bandwidth a range of FREQUENCIES in a section of the electromagnetic spectrum that is expressed in terms of the lowest and highest signal in a FREQUENCY (ENGINEERING) band. For example, the bandwidth or FREQUENCY range of VHF is from 30 to 300 MEGAHERTZ (MHZ). Bandwidths are said to be narrow or wide, depending on the difference between the highest and lowest frequency in them.

barn doors metal appurtenances that are attached to lighting instruments and act as extensions of them. They consist of four rectangular flaps, one on each side and on the top and bottom of the instrument. They can be adjusted, like a shutter, to focus and direct the beam of light and cut it off where it is not desired.

barter syndication the method whereby programs are traded for COMMERCIAL TIME. The shows are offered free of charge to stations by SYNDICATORS, with commercials for different products provided by a national advertiser. The station gains a program with no cash outlay. This technique of syndication is sometimes known as "advertiser-supported syndication." The system has

created an important marketplace for original FIRST-RUN SYNDICATED PROGRAMS.

There are three major types of barter syndication: (1) FULL BARTER in which the national advertiser retains all of the commercial time but the station receives the program free of charge; (2) CASH/BARTER in which the station pays a small license fee for the program and trades commercial time for the balance; and (3) SPLIT BARTER, an arrangement whereby the station pays no license fee but retains some of the time for its own sale.

Bartlesville pay-cable experiment a 1957 PAY-TV experiment that occurred in Bartlesville, Oklahoma. It became the first extensive PAY (PREMIUM) CABLE SERVICE system in the nation. The operation was discontinued the following spring.

base lighting See FILL LIGHTING.

basic cable service the primary part of a cable system's offerings to customers. In the early days of cable, it denoted the programming that a subscriber received for a single monthly fee. At that time, basic service consisted of the retransmission of local stations, the importation of one or two DISTANT SIGNALS, and a community event/activities wheel or weather and bulletin board.

Today, the many basic cable channels are often difficult to distinguish from the PAY (PREMIUM) CABLE SERVICE channels offering similar programs. Most basic channels, however, have COMMERCIALS while the pay channels do not carry advertising.

below-the-line costs a part of a method of financial accounting for television, cable, and video production. The term refers to the placement on an accounting sheet of all the technical costs incurred in developing a show or series. Charges for engineering equipment, props, and rental of special gear or effects are placed in this category in the budget. They are literally placed below a line on an accounting page, distinguishing them from ABOVE-THE-LINE COSTS.

best boy the chief assistant to the GAFFER on a motion picture set; used in television in the production of MINISERIES and MADE-FOR-TV MOVIES. The best boy serves as the right hand to the gaffer, is in charge of the lighting crews and is responsible for the acquisition of all lighting equipment. The term came into use when all stagehands were men. It may have originated to recognize the best (and perhaps the most eager) gofer who could also organize and control other crew members.

best time available (BTA) the designation on the written order for COMMERCIAL TIME by an ADVERTISING AGENCY where the scheduling of a SPOT is determined by the cable or television operation. Such an arrangement is less expensive than buying predetermined or fixed time periods, but the scheduling of a spot is at the discretion of the television or cable operation.

Beta format a videotape format developed by the SONY CORPORATION. The Beta format introduced the U.S. public to home video and TIME SHIFTING in the fall of 1975. An extensive and successful advertising CAMPAIGN made the term "Betamax" almost synonymous with videotape recording in consumers' minds.

Consumer interest in the machine was enhanced by publicity surrounding the unsuccessful suit by Universal Studios to stop Betamax sales by Sony in the so-called BETAMAX CASE.

In 1977 MATSUSHITA introduced a slightly larger and incompatible half-inch VHS videocassette machine. For a period of five years, the two VIDEOTAPE FORMATS waged a battle over prices and length of recording time. Most technical experts cited the Beta format as superior in quality. The VHS machine, however, gained greater acceptance with the public, and although some diehard Betaphiles continue to extol the virtues of the technology, the VHS format had become the *de facto* home videotape recording standard by the late 1980s. No new standard home video Beta machines are now being sold in the United States and it is increasingly difficult to obtain prerecorded videocassettes for the existing machines. The format, however, continues to exist in the professional BETACAM FORMAT camcorders, which are the mainstay of ENG operations in the broadcast industry.

Betacam format the most popular professional, portable ENG unit in the United States. This CAMCORDER can record some 20 minutes on a special BETA format videocassette. The camcorder cannot play back, however, and the tapes are not compatible with conventional Beta decks.

Betamax case litigation that established the legitimacy of home video recording in the United States. The case decided whether the use of videocassette machines to record programs on the viewer's home television set violated federal COPYRIGHT laws. In 1979, the Hollywood plaintiffs argued that in manufacturing the machine, the Sony Corporation knew the device would be used to infringe on their copyrights of televised programs and motion pictures. In 1984, the Supreme Court held that recording at home was not a violation of the copyright law, finding that most of the recording was for TIME-SHIFT purposes and the convenience of the public. The court said that programs recorded for this reason and not for resale fell under the fair-use provisions of the law.

bicycling the shipment of videotape recordings of television programs from one transmitting entity to another. In order to save film and videotape costs, the programs are sent to a station or cable system when they are needed for the station's schedule. After a program is broadcast, that station or system sends it on to the next operation.

bidirectional microphone a type of mike that is sensitive to sounds from front and back, but not from the sides. Sounds coming in from the sides cancel one another out, leaving only the front and rear sounds. The mike is sometimes called a "figure eight" or a "pressure gradient."

billboards animated or still graphics that depict the sponsors' logos and are usually KEYED over the opening shots of a program. Billboards are commonly used to identify all of the PARTICIPATING SPONSORS of a televised event.

billings the total amount of money charged to the client by an ADVERTISING AGENCY, including all costs for the purchase of COMMERCIAL TIME, production and talent costs, and research fees.

biomedical communications the use of AUDIOVISUAL COMMUNICATIONS devices and methods in healthcare situations. Teaching centers associated with university and Veterans Administration (VA) hospitals use many types of media, as do school health centers, clinics and private physicians' offices.

bit a show-business term that refers to a short piece of comedic business in a sketch or a routine in a stage or television show. It may consist of a brief exchange with another performer or an abbreviated and succinct solo phrase, gesture, action, or sound.

Black Awareness in Television (BAIT) a membership organization that concentrates on producing black programs for television, radio, video, and film. It produces public affairs programs and promotes the visibility of consumer products produced by blacks.

Black-Owned Broadcasters Association See NATIONAL ASSOCIATION OF BLACK-OWNED BROADCASTERS (NABOB).

blacklisting a technique employed by the political right in the 1950s that pressured the networks to refuse to employ any person suspected of being a Communist, or who was a "leftist" or "fellow traveler."

Watchdog organizations compiled lists that contained the names of "pro-Communists" or "people with subversive ideas" or those with ultra-liberal tendencies. A producer, writer, director, or actor whose name appeared on a clandestine list was often quietly fired and later became unemployable.

No one usually acknowledged the existence of the blacklists, but one public and highly visible list surfaced in the form of the book *Red Channels,* published in 1950. The movement reached its peak with the posturings of Senator Joseph McCarthy in 1954. His attempt to root out subversives in the Army and the federal government was exposed as a terrible sham in the ARMY–MCCARTHY HEARINGS. Blacklisting was finally ended with the resolution of the Faulk case in 1964. (See also MCCARTHYISM.)

blanking the brief period during the SCANNING LINES process in which the video signal is suppressed. The electron beam in a television CATH-ODE RAY TUBE (CRT) travels constantly across and up-and-down the screen, but there is a regular period when the scanning beam returns from right to left and from bottom to top. This brief pause is known respectively as horizontal and vertical blanking, and during this period, the video signal is suppressed and invisible.

block a procedure that designs the movement of performers and cameras during the course of the rehearsal of a television program. The term was borrowed from the theater and it indicates the design of the predetermined actions in a production.

block programming a technique of scheduling similar television programs on a station, network, or cable system within a relatively brief period of time. The programs, which have a common appeal, are scheduled during a two- or three-hour block in a DAYPART.

blockbuster movie an enormously successful "A" TITLE in home video that has also been a megahit in theatrical release or on television. These outstanding films are sometimes LOSS LEADERS when released on television or home video because they fail to earn back their cost, but they attract an audience or customers that can be directed to other shows or titles.

Blockbuster Video the largest VIDEO RETAIL CHAIN in the United States. The Florida-based firm operates mammoth VIDEO SUPERSTORES throughout the nation and world, which are known for the BREADTH AND DEPTH of their stock. In 1993, the company announced plans to merge with VIACOM ENTERPRISES.

blooming effect   a technical aberration in a television picture. The defect consists of a very small distortion in the size of an image. It resembles a halo and the picture is also slightly out of focus.

Blue Book   a document that was the first formal statement of the program standards expected of broadcast stations licensed by the FEDERAL COMMUNICATIONS COMMISSION (FCC). Officially titled "Public Service Responsibility of Broadcast Licensees," it was issued on March 7, 1946. The mimeographed report derived its name from its deep blue cover. Although it was issued in the era of radio, its principles have never been withdrawn by the Commission and could conceivably apply in the days of television.

board it up   an ADVERTISING AGENCY slang term that is a directive for an artist to sketch out an idea for a COMMERCIAL on poster board. To "board it up" implies that the concept is regarded seriously by the agency.

bookends   an innovative type of television COMMERCIAL developed in the 1980s. The concept usually consists of two 15-second SPOT (ANNOUNCEMENTS) for one product, separated by one or more commer-

cials for products from other advertisers. The two spots are usually at the beginning and end of a commercial period.

books   softcover publications that list the RATINGS, SHARES and other audience research data for radio, television and cable operations. They are periodically published by A. C. NIELSEN and the ARBITRON COMPANY and issued to subscribers of their audience measurement services. (See also POCKETPIECE (PP) reports.)

boom   an adjustable and portable stand for a microphone that is used in both studio and REMOTE television productions. It supports a mike and keeps it out of sight of the camera, usually above and in front of the performer.

booster station   a low-power repeater of a full-power television station that simply amplifies the signal of the parent station and rebroadcasts it on the same channel to an immediate area. Boosters always broadcast on the same channel as the parent and thus differ from TRANSLATOR STATIONS that convert an incoming signal from a parent station and rebroadcast it on another channel.

boutique agency   a small ADVERTISING AGENCY that specializes in particular aspects of an advertising CAMPAIGN, such as the creation of art or copy for the production of commercials. As opposed to a FULL-SERVICE ADVERTISING AGENCY, a boutique concentrates only on very specific, high quality media services.

box house a large-volume, high-discount dealer that sells television sets, videocassette recorders (VCR), and other electronic gear to consumers from a large, no-frills store. The term is derived from the practice by the retailer of selling the devices in the manufacturer's original packing box.

brainstorming a creative procedure in advertising that is designed to elicit all kinds of spontaneous ideas from a group seeking to develop a COMMERCIAL or an advertisement.

Bravo a PAY (PREMIUM) CABLE SERVICE network that offers international films and performing arts programs. The films, which vary between English-language and dubbed foreign-language versions, are introduced by a host who provides background information. Bravo is a part of RAINBOW PROGRAM ENTERPRISES, which is a subsidiary of CABLEVISION SYSTEMS CORPORATION (CVC).

breadth and depth the extent of the number of titles (breadth) and the number of copies of those titles (depth) in a video retail store.

bridging amplifier a device that boosts cable television signals. Bridging amplifiers amplify the signals in a system and send them on.

British Academy of Film and Television Arts (BAFTA) a nonprofit organization that awards the British equivalent of EMMYS and OSCARS in the United Kingdom. The academy is that country's combined equivalent of the ACADEMY OF MOTION PICTURE ARTS AND SCIENCES (AMPAS), the NATIONAL ACADEMY OF TELEVISION ARTS AND SCIENCES (NATAS), and the ACADEMY OF TELEVISION ARTS AND SCIENCES (ATAS) in the United States.

British Broadcasting Corporation (BBC) the public broadcasting organization of the United Kingdom, established in 1927. It operates independently of the government and receives most of its revenues from annual taxes paid to the Post Office by the owners of radio and television sets. There are no COMMERCIALS on the BBC's radio or television networks.

The corporation operates four radio networks and local radio stations in the country's major cities. It also operates a worldwide radio broadcasting service on shortwave as well as two television networks called BBC-1 and BBC-2.

British Sky Broadcasting a DIRECT BROADCAST SATELLITE (DBS) organization formed in late 1990 as the result of a merger of British Satellite Broadcasting and Sky Television. It serves the United Kingdom and some European countries with English-language programming. The merged operation retains the "Sky" in the corporate name to promote sports, movies, news, and entertainment channels.

British Universities Film and Video Council Ltd. (BUFVC) an organization that develops and coordinates the use of film, video and other AU-DIOVISUAL COMMUNICATIONS devices in British universities.

broadband communication (systems) systems that deliver multiple channels over a wide BANDWIDTH to users or subscribers. Cable television (CATV) is the quintessential broadband communications system.

Broadcast Advertising Reports (BAR) a New York organization that monitors television commercials in more than 75 markets. The company publishes periodic reports about the airing of commercials for particular products. ADVERTISING AGENCIES subscribe to the reports to check on commercial activity.

Broadcast Cable Financial Management Association (BCFMA) a nonprofit organization consisting of comptrollers, vice presidents of business affairs, and financial managers in the broadcasting industry.

Broadcast Designers Association (BDA) a nonprofit membership organization that includes graphic artists, art directors, and others involved in images and graphics for television and cable operations.

Broadcast Education Association (BEA) a nonprofit organization of graduate students and professors that is involved in instruction and research in radio, television, and related communications technology.

Broadcast Financial Management Association (BFMA) See BROAD-CAST CABLE FINANCIAL MANAGEMENT ASSOCIATION (BCFMA).

Broadcast Music Inc. (BMI) a nonprofit performing-rights organization that licenses the performance rights of copyrighted music. BMI issues licenses to television stations, cable systems, and video companies, and collects fees for its member composers, lyricists, and publishers.

Broadcast Pioneers a nonprofit society consisting of individuals who have been in the broadcasting and cable industry for more than 20 years. It includes many retired executives.

Broadcast Pioneers Library a library that contains a wealth of information on the history of broadcasting. It is a repository for oral history tapes, transcripts, disc and wire recordings, early research studies, books, scripts, and personal documents. It is housed at the NATIONAL ASSOCIATION OF BROADCASTERS (NAB) headquarters in Washington, D.C.

Broadcast Promotion and Marketing Executives Inc. (BPME) a national association of promotion and public relations personnel in radio/television and cable. The group changed its name to Promax (a dirivative acronym for Program Executives in the Electronic Media) in 1993, but will use both names for an unspecified period of time.

Broadcast Rating Council (BRC) See ELECTRONIC MEDIA RATING COUNCIL (EMRC).

Broadcast Television in a Multi-channel Marketplace See PEPPER PAPER.

Broadcast Traffic and Residuals (BT&R) a firm handling talent payments and trafficking functions for television and video production. BT&R handles all the cumbersome and time-consuming administrative paperwork in dealing with performers belonging to the SCREEN ACTORS GUILD (SAG) and the AMERICAN FEDERATION OF TELEVISION AND RADIO ARTISTS (AFTRA).

broadcast(ing) one of the mass media, which involves the dissemination of information by radio waves or electromagnetic radiation through the air for the purposes of communications. The FEDERAL COMMUNICATIONS COMMISSION (FCC) adds a further important caveat to the definition in specifying that the transmissions must be for reception by the general public.

The term encompasses both radio and television transmissions, since television is a form of radio involving the simultaneous distribution of both sound (AUDIO) and picture (VIDEO). The word broadcasting is not used to describe transmissions via point-to-point MICROWAVE RELAY systems (or even point-to-multipoint systems such as SATELLITE transmissions to TELEVISION RECEIVE ONLY [TVRO] dishes), because the purpose of such transmissions is for private, not general, public reception. Further, the term is not used in describing cable systems inasmuch as transmission by that technology is through COAXIAL CABLE rather than through the air.

The word stems from the manner by which a farmer casts seed over a broad area in planting some crops. It evolved into the language as a noun, a verb (to broadcast), and an adjective (broadcasting station).

broads lighting instruments designed to generally illuminate a television set in a studio by throwing soft light over a large area.

budget video low-priced, prerecorded videocassettes offered for sale in convenience stores, supermarkets and other mass market outlets. Usually distributed by RACK JOBBERS, the cassettes are often PUBLIC DOMAIN films.

bumper a brief announcement in television programming designed to separate the program content from a COMMERCIAL.

bumping See DUBBING.

burn a slang expression referring to the image that remains on a camera tube when it has been focused on a subject too long or exposed to a bright light.

buying groups organizations within the home video industry that obtain volume discounts for prerecorded video titles from WHOLESALERS for their retail store members. The organizations also provide their membership with other services that may include reduced charges on long-distance phone calls, group insurance, low prices for candy and other ancillary items, and similar group benefits.

buyout a compensation practice by which the talent in a prerecorded COMMERCIAL or program receives a one-time payment and waives the rights to remuneration for all future transmissions of that appearance. Buyouts are distinct from RESIDUALS, which pay the performer each time a commercial or program is aired.

C

C–band satellites relatively low-powered communication SATELLITES that utilize the C-band and cover the entire United States. They are used by cable systems and television broadcasters to receive the signals on large TELEVISION RECEIVE ONLY (TVRO) dishes. Transponders aboard the satellite DOWNLINK the signal using FREQUENCIES between 3.7 and 4.2 GIGAHERTZ (GHZ).

C–clamp a metal device shaped like the letter C, used to connect lighting instruments to a pipe grid above a television studio. A version of the C-clamp is also used to temporarily hold flats together or pieces of scenery in place.

C–mount a type of connection used to attach a lens to a television or film camera. The connection is standardized so the threads, the hole, and the base are all compatible. Nearly all 16mm cameras and most television cameras use the ubiquitous C-mounts.

C–Span See CABLE SATELLITE PUBLIC AFFAIRS NETWORK.

Cable Alliance for Education See CABLE IN THE CLASSROOM.

Cable Communications Policy Act of 1984 federal legislation that attempted to establish a "national policy concerning cable communications." The power of the COMMUNICATIONS ACT OF 1934 and the authority of the FEDERAL COMMUNICATIONS COMMISSION (FCC) versus the authority of state and local agencies to regulate cable had been unclear.

In October 1984 Congress passed the Cable Communications Policy Act, which became Title VI of the Communications Act of 1934. The law established policies for franchises and renewals, piracy, cable rate regulation, channel usage, ownership, and EQUAL EMPLOYMENT OPPORTUNITY (EEO). It also established the jurisdictional responsibilities among federal, state, and local authorities for cable television.

cable compatible a consumer electronics phrase coined in the 1980s to describe television sets and videocassette recorders that are designed to be directly connected to a CABLE DROP in a home. The units (sometimes called "cable-ready sets") contain a tuner that can receive all cable as well as all broadcast channels.

cable drop   the last connecting element of a cable system in a TREE NETWORK configuration. The cable drop consists of a small COAXIAL CABLE (about one-quarter of an inch in diameter) that connects the FEEDER CABLE of the distribution system to the subscriber's home and then to his CONVERTER or television set.

Cable in the Classroom   an organization of MULTIPLE SYSTEM OPERATORS (MSO) and cable programming networks. The group is dedicated to connecting cable service to every school and to providing programming for use in the K–12 curriculum. It was founded in 1990 as the Cable Alliance for Education.

Cable News Network (CNN)   a BASIC CABLE SERVICE that became the world's first 24-hour channel devoted entirely to news. Inaugurated in June 1980, CNN has become the channel to watch for breaking news in times of crisis.

cable registration   procedures for obtaining a CERTIFICATE OF COMPLIANCE. To register, a cable television operator must send the following information to the FCC: (1) the legal name of the operator, (2) the assumed name (if any) used for doing business in the community, (3) the mailing address, (4) the date the system will provide services to subscribers, (5) the name of the community or area served, (6) the television broadcast signals to be carried that have not previously been certified or registered, and (7) a statement of the proposed EQUAL EMPLOYMENT OPPORTUNITY program. A cable system may commence operations immediately upon filing the registration.

Cable Satellite Public Affairs Network (C–Span)   private, nonprofit cooperative cable networks, financed by affiliate fees. C–Span consists of two cable networks that provide live coverage of congressional hearings and public events in the United States and abroad on a 24-hour-a-day basis.

cable signal-leakage requirements   part of FEDERAL COMMUNICATIONS COMMISSION (FCC) regulations that place restrictions on electronic signal-leakage in cable systems. They are intended to prevent interference by cable operators with radio FREQUENCY users, particularly users of restricted frequencies in the aeronautical BANDWIDTHS.

cable spot advertising   the sale of commercial SPOTS by cable companies to advertisers.

cable television (definition)   defined by Section 602(6) of the CABLE COMMUNICATIONS POLICY ACT OF 1984 as "a facility consisting of a set of closed circuit transmission paths and associated signal generation, reception, and control equipment that is designed to provide cable service, which includes video programming and which is provided to multiple members within a community."

Section 602(5) of the Act defines "cable service" as "(A) the one-way transmission to subscribers of (i) video programming or (ii) other programming services and (B) subscriber interaction, if any, that is required for the selection of such video programming or other programming service."

According to the FCC in 1990, only video delivery systems that use cable, wire, or other physically closed/shielded transmission paths to serve subscribers are considered cable systems.

Cable Television Administration and Marketing Society (CTAM) a nonprofit TRADE ASSOCIATION that acts as a venue for the exchange of information and ideas among marketing and sales management personnel in the cable industry.

Cable Television Report and Order of 1972 regulations issued for cable television by the FEDERAL COMMUNICATIONS COMMISSION (FCC). The rules as adopted in 1972 required cable television operators to obtain a CERTIFICATE OF COMPLIANCE from the FCC prior to operating a cable system or adding a television broadcast signal. Other rules concerned EQUAL EMPLOYMENT OPPORTUNITY (EEO), SYNDICATED EXCLUSIVITY, and FRANCHISING. The systems that originated programming were subject to FAIRNESS DOCTRINE and EQUAL TIME (OPPORTUNITY) RULES similar to those

that covered broadcast operations. The FCC eliminated or modified many of the rules during the next 12 years. The other rules remained in effect, however, until the passage and implementation of the CABLE COMMUNICATIONS POLICY ACT OF 1984, which superseded all the 1972 rules.

cable theft rules Section 633 of the CABLE COMMUNICATIONS POLICY ACT OF 1984 that provides for damages and penalties for the unauthorized use of cable services. It allows for specific criminal and civil remedies for cable theft but makes a distinction between people engaged in the practice for personal gain and those who are involved for commercial advantage. A first-time offender may be fined up to $25,000 and a previously convicted individual up to $50,000 with imprisonment for two years.

cablecast See TELECAST.

CableLabs an organization that serves as a technical research and development consortium of cable television system operators.

Cabletelevision Advertising Bureau (CAB) an organization that promotes and tracks advertising on BASIC CABLE SERVICE networks, MULTIPLE SYSTEM OPERATORS (MSO), INTERCONNECTS, and local systems. The group conducts extensive research about advertising on cable and encourages its growth with press releases, promotional activities and seminars.

Cablevision Systems Corporation (CVC) one of the nation's largest cable MULTIPLE SYSTEM OPERATORS (MSO). Cablevision and its affiliates serve subscribers in 11 states. It operates systems ranging in size from those with fewer than 5,000 subscribers to the largest single cable television system in the United States—the Long Island system (which has more than 525,000 subscribers). Cablevision also owns RAINBOW PROGRAM ENTERPRISES, which produces and distributes BRAVO (a pay cable service dedicated to international films and the performing arts) and American Movie Classics (offering Hollywood films from the 1930s to the 1970s).

In addition the company operates 10 regional sports channels and Sportschannel America. Cablevision and NBC are involved in a joint cable programming venture between the company's programming entities and NBC's CONSUMER NEWS AND BUSINESS CHANNEL (CNBC).

California revolution, the an irreverent and absurd school of advertising that was pioneered by the ADVERTISING AGENCY Chiat/Day in the 1980s. The CAMPAIGNS developed by the Venice, California agency changed the perception of many viewers about commercials. Chiat/Day simply ignored rules, relying on an unexpected iconoclastic send-up approach with striking images designed to stir the emotions, such as a chair shedding its upholstery to the music of "The Stripper," to point out the listings of furniture strippers in the *Yellow Pages*.

call letters combinations of alphabet characters that are used to identify radio and television broadcasting stations. In the United States, call letters are assigned by the FEDERAL COMMUNICATIONS COMMISSION (FCC). The FCC requires that stations identify themselves at the beginning and end of programming every day and periodically during their broadcast periods. The use of call letters distinguishes one station from another in the crowded broadcast spectrum and helps avoid confusion among the stations and the audience.

Full-power broadcast television stations in the United States may request from the FCC any combination of letters beginning with W or K that is not in use or sounds similar to those of other stations' signs. (See also COMMERCIAL, STATION BREAKS, and STATION IDENTIFICATION [ID].)

callback the second phase of the auditioning process in which a performer is called back to perform or AUDITION again or for a discussion of the role.

camcorder videotape recording units and television cameras in one package. Battery-operated, they are lightweight and portable. Professional versions are used for EFP and ENG operations. (See also COMPONENT VIDEO SYSTEM RECORDING.)

camera angle the perspective from which a television or film camera photographs a scene or subject. Camera angles are usually measured from the camera operator's eyepiece. From that point there can be high or low angles or angles from the right or left of a subject. (See also FRAMING.)

campaign an overall advertising plan designed for a specific client by an advertising agency. Borrowed from military jargon, the term describes a carefully constructed and orchestrated series of advertising elements, which are related to one another and scheduled over a defined period of time. Television commercials, radio spots, print ads, and billboards support each other and have a cumulative effect on the audience. (See also MEDIA PLAN.)

Canadian Association of Broadcasters (CAB) a nonprofit association of radio and television stations in Canada. The CAB was formed to promote and defend its members' interests in all aspects of private radio and television broadcasting.

Canadian Association of Motion Picture and Electronic Recording Artists (CAMERA) a trade union that represents directors of photography, camera operators, and first and second assistant camera operators in negotiations with producers of feature films, television programs, COMMERCIALS, and DOCUMENTARIES.

Canadian Broadcasting Corporation (CBC) a publicly owned corporation, established in 1936 by an Act of Parliament. The CBC provides a national radio and television broadcasting service in Canada's two official languages, English and French.

It is financed mainly by public funds voted annually by Parliament and supplementary revenue is obtained from commercial advertising on the CBC television networks. CBC Radio is virtually free of commercial advertising.

The CBC operates seven national services including an English television network, a French television network, and the National Satellite Channel, which delivers to Canadians the proceedings of the House of Commons via satellite and cable.

Canadian Cable Television Association (CCTA) a nonprofit association that represents the industry to the public and promotes standards of excellence and codes of conduct in the field.

Canadian Department of Communications a department of the Canadian government that ensures the orderly operation of Canada's communications systems. It also protects the freedom of all citizens to choose a wide selection of Canadian cultural products and information services.

The department promotes the development and use of a national communications system that links all regions through a variety of conventional and new technologies including television, telephone, SATELLITE, electronic media, radio, and FIBER OPTICS. The Minister of Communications is responsible to the nation's Parliament for enabling legislation, regulatory agencies, and branches of government including the CANADIAN RADIO–TELEVISION AND TELECOMMU-

NICATIONS COMMISSION (CRTC), the CANADIAN BROADCASTING CORPORATION (CBC), and the NATIONAL FILM BOARD OF CANADA (NFB).

Canadian Film Editors Guild (CFEG) See DIRECTORS GUILD OF CANADA (DGC).

Canadian Radio-Television and Telecommunications Commission (CRTC) an independent agency that regulates two main areas: broadcasting and telecommunications. The CRTC is comparable to the FEDERAL COMMUNICATIONS COMMISSION (FCC) in the United States. It was established to regulate and supervise all sectors of Canada's broadcasting system including AM and FM radio, television, cable, pay-TV, and specialty services. The CRTC grants, amends, or renews licenses; monitors the performance of licensees; and establishes broadcasting regulations and policies. In 1975 another law assigned CRTC the responsibility for the regulation of telecommunications (telephone) activities, mainly with respect to rates and terms of service.

cans headphones worn by television and radio production personnel. The slang term is also used to refer to the circular metal containers in which a film is stored, leading to the industry use of the phrase, "in the can," for a completed motion picture or television program.

capacitance electronic disc (CED) one of the casualties of the videodisc format war in the early 1980s. Along with a similar system, the VIDEO HIGH DENSITY (VHD) VIDEODISC format, the CED was made obsolete by the better quality LASER VIDEODISC (LV) technology.

The CED technology was developed in the 1960s and 1970s and introduced in 1981 by the RADIO CORPORATION OF AMERICA (RCA) under the name SelectaVision.

Similar to an audio turntable, the CED machine used a miniature stylus that physically contacted the disc and deciphered the encoded electronic information, translating it into pictures and sound on the TV set.

The even lower-priced VIDEOCASSETTE technology, which could record as well as play back, however, came to dominate the consumer electronics industry during the early 1980s. RCA withdrew the CED machine from the market in 1984.

Capital Cities/ABC Inc. a megacompany that consists of the ABC TELEVISION NETWORK, eight television stations, seven radio networks and 21 radio stations. The company also publishes nine daily newspapers, numerous weekly newspapers and shopping guides, and various periodicals and books. In addition the firm is a supplier of programming to the cable industry with partnerships that operate ESPN, the ARTS & ENTERTAINMENT NETWORK, and LIFETIME. The Video Enterprises Division of the company licenses programming to domestic and international home video markets and to television stations abroad.

cardioid microphone a mike named after its pickup pattern, which is shaped like a heart. The device is unidirectional and is sensitive to sounds from both sides but especially from the front.

Carnegie Commission on Educational Television (Carnegie I) a commission established in 1965 by the Carnegie Corporation to study and make recommendations regarding the future of noncommercial television in the United States. After more than a year of study, the 15-member commission issued its report, "Public Television: A Program for Action" in 1967. It recommended a name change from "educational" to "public" television, and the creation of a Corporation for Public Television (later the CORPORATION FOR PUBLIC BROADCASTING [CPB]), which would receive and distribute funds from the federal government and foundations. The report was widely read and well received and many of the recommendations were incorporated into the PUBLIC BROADCASTING ACT of 1967.

Carnegie Commission on the Future of Public Broadcasting (Carnegie II) a second commission formed to study the impact and future of public broadcasting. After a year-and-a-half, the new commission (with new members) issued its report titled "A Public Trust" in 1979. The report recommended a restructuring of the industry and a massive increase in federal funding. Unlike the first report, Carnegie II's recommendations did not have much impact on the noncommercial system.

CARS an acronym that stands for community antenna relay service. Authorized by the FEDERAL COMMUNICATIONS COMMISSION (FCC), this MICROWAVE RELAY system acts as a cable relay service to transmit signals via microwave for local distribution, inner-city relay, and remote television pickup.

Cartrivision a VIDEOTAPE FORMAT that competed with its predecessor, the CBS-developed EVR and the 3/4-INCH U (EIAJ) VIDEO RECORDING format created by the SONY CORPORATION. The first Cartrivision units went on sale in Sears stores in the Chicago area in June of 1972 after only two years of development. The device was sold only as a console model, which included the videocassette machine and a color TV set, all for some $1,600. Developed by Cartrivision Television Inc. and manufactured by Avco, the machine used half-inch tape and could record programs as well as play them back. The picture quality of the playback tape was not good and sales were not as expected. The firm lost money steadily and, in spite of layoffs and a massive reduction in operating expenses, it had to declare bankruptcy in June 1973. The format did not survive.

cash/barter syndication a type of syndication that involves elements of FULL-BARTER SYNDICATION and CASH SYNDICATION in the sale of television programming. In this transaction the local station or cable operation pays a lower cash fee to license the program but gives up some of the available COMMERCIAL TIME within the show to national advertisers. The local operation retains the remainder of the time for its sale.

cash syndication   a method of syndication in which a television or cable operation pays the SYNDICATOR (distributor) a flat fee to license a program for transmission. In this transaction the local station or system purchasing the right to air the program(s) a number of times over a specified period. The operation assumes all of the risk in purchasing the show and must sell the COMMERCIAL TIME locally or SPOTS nationally. It is the simplest and most straightforward system of syndication and is the largest segment of the business. (See also CASH/BARTER SYNDICATION.)

Cassandra   a very sophisticated computer software system used by A. C. NIELSEN, which provides information about selected programs that have been previously transmitted on a market-by-market basis. Program data, including RATINGS and DEMOGRAPHICS from past years, can be compared and historically analyzed with current programs in a variety of ways to help determine trends and assist in buying COMMERCIAL TIME.

catalog product   titles listed in the catalog or sales list of a PROGRAM SUPPLIER in the prerecorded video industry. The catalog titles usually sell regularly and steadily throughout the year and are the equivalent of the backlist in the book industry.

cathode ray tube (CRT)   one version of an electronic vacuum tube designed for the display of images. There are two types of cathode ray television tubes, one for black-and-white television and one for color.

Catholic Broadcasters Association
See UNDA–USA.

Catholic Conference of Broadcasting   the communications department of the U.S. Catholic Conference. The group addresses the church's apostolate via the print and electronic media and provides advice and technical assistance for radio and television network programs.

CAV (constant angular velocity)
See LASER VIDEODISC (LV).

CBS Inc.   One of the major commercial full-service national television networks. The company began in early 1927 and by 1929, the fledgling network was losing money and was purchased by William S. Paley, who was one of America's leading broadcasting pioneers. He simplified the name to the Columbia Broadcasting System (CBS) and this was shortened to CBS Inc. in 1974. The network's image was enhanced during World War II by its outstanding news operation, headed by Edward R. Murrow.

Capitalizing on the success of the radio network, Paley developed the television operation into one of the most consistently successful companies of its kind in the world. CBS became the number one network in ratings in 1955 and maintained that position for 21 years.

The Loews Corporation, headed by Lawrence A. Tisch, acquired a majority interest in the company in 1986, but Paley was asked to remain. He served as chairman until his death in 1990 and was succeeded by Tisch.

In addition to the TV network of 212 AFFILIATED STATIONS, CBS owns five television stations (serving New York, Chicago, Los Angeles, Philadelphia and Miami), the CBS radio network, and 20 radio stations.

CD + G initials that stand for compact disc + graphics. It is a standard musical compact disc that also contains graphics such as the lyrics to a song or other related textual material. The visual material is encoded in some of the empty space on the CD and can only be displayed on a television set by using a special player.

CD-I initials that stand for compact disc, interactive. Announced by the Philips Corporation in March 1987 and developed in conjunction with the SONY CORPORATION, the technology is designed to look like a CD player and to operate as simply as a video game, using a standard TV set. It has great potential in the educational field and in electronic publishing. The machine will play back discs that contain pages of text, still video frames, graphics, and CD-quality audio. The specifications for the device have been adopted by other manufacturers, making any programs developed for use on CD-I compatible with any CD-I player, worldwide.

CD-ROM initials that stand for compact disc read-only memory (pronounced cee-dee-rom). The device was the first video format to evolve from the audio COMPACT DISC (CD). The machines were introduced in 1985 by Philips and the SONY CORPORATION as data-storage peripherals to the personal computer (PC). CD-ROM encoded discs are permanent collections of textual information using a DIGITAL COMMUNICATIONS process. The five-inch discs hold the equivalent of some 150,000 printed pages or about 1,000 times more data than a PC floppy disc. The discs are inserted into the CD-ROM player, which can then be controlled by the PC to afford random access to the data. Users cannot change the data (hence the designation read only) but a new development titled WORM (write once, read many times) allows for the user's creation of data. Most consumer PCs in the future will have built-in CD-ROM players.

CDTV a COMPACT DISC (CD) format that is commercially competitive with the CD-I format. The two types of discs are incompatible; a disc in one format will not play back on the other's machine. An enhanced version of the CD-ROM format, the

CDTV disc allows the storage of animation, sound, and text. Both discs may be superseded in the late 1990s by the DVI technology, which will feature even more interactivity because of its use of DIGITAL VIDEO COMPRESSION.

Ceefax   a one-way broadcast teletext system developed by the BRITISH BROADCASTING CORPORATION (BBC) in the late 1960s. Its name is a corruption of the phrase "see facts." Ceefax transmits some 100 pages of information such as news, weather, and entertainment options, which can be seen on the vertical BLANKING interval of the TV set. The information can be called up at any time by the viewer using a keypad and a decoder.

censorship   the practice of suppressing anything perceived as objectionable. The freedom of speech and the press, as guaranteed by the First Amendment, was specifically expanded to broadcasting in the COMMUNICATIONS ACT OF 1934. Section 326 of that Act states:

> Nothing in this Act shall be understood or construed to give the Commission the power of censorship over the radio communications or signals transmitted by any radio station, and no regulation or condition shall be promulgated or fixed by the Commission which shall interfere with the right of free speech by means of radio communication.

"Radio communication" was defined in the Act as the "transmission by radio of wire signs, signals, *pictures* [emphasis added], and sounds of all kinds . . ." thus allowing for the legislation to cover television many years later. While the Act prevents the FEDERAL COMMUNICATIONS COMMISSION (FCC) from censoring programs, it does not prohibit private companies or individuals from making editorial and content decisions about broadcast programs.

Center for Communication   a nonprofit organization that brings faculty and students of communications together in dialogue with practitioners in the industry. Founded in 1980, it sponsors seminars and panel discussions on television, cable, and the newer technologies. It is supported by foundations and gifts.

Center for New Television (CNTV)   a nonprofit organization serving as a resource center for video and community media producers, an activist group for the invention of new formats and an advocate for independent artists. The center offers low-cost access to video production and POSTPRODUCTION equipment, technical assistance to not-for-profit and grassroots community organizations, and support in fundraising and grant management. (See also ALTERNATIVE TELEVISION.)

Central Educational Network (CEN)   a private, nonprofit, regional PUBLIC TELEVISION (PTV) network providing INSTRUCTIONAL TELEVISION (ITV), postsecondary education, and general audience programming to its many stations. CEN primarily serves member stations in the Midwest.

certificate of compliance   a formal requirement to force cable companies to adhere to some federal standards. It was incorporated in the CABLE TELEVISION REPORT AND ORDER OF 1972 by the FEDERAL COMMUNICATIONS COMMISSION (FCC). The purpose was to impose guidelines in the franchising process and require companies that had been awarded a FRANCHISE to comply with federal standards. The certificate was abolished by the FCC in 1978 and replaced by a CABLE REGISTRATION requirement.

channel   a FREQUENCY band allocated by the FEDERAL COMMUNICATIONS COMMISSION (FCC) for the transmission of a signal. U.S. standards require a 6-MEGAHERTZ (MHZ) bandwidth channel. To avoid electronic interference between channels, the FCC assigns channels geographically and specifies certain frequencies for use by AM and FM radio and UHF and VHF television stations.

channel realignment rules   Section 625 of the CABLE COMMUNICATIONS POLICY ACT OF 1984, which gave cable systems great freedom in selecting where channels could be located on their systems.

The rules give the operator freedom to "rearrange, replace, or remove a particular cable service" and to "rearrange a particular service from one service tier to another." There are some conditions that must be met and PEG CHANNELS must be carried, but in the main, cable operators are free to switch channel posi-

tions upon 30 days' notice to the FRANCHISE authority.

character generator (CG)   an electronic image device that generates a sequence of signals that forms words and symbols on a TV screen. It allows production personnel to develop titles for a program or to place captions within televised images. The unit resembles a computer terminal keyboard and the operator types and stores words, phrases, names, or captions on pages that can be called up for SUPERIMPOSITION or KEYING over the picture when the unit is connected to the video system. In Europe the device is often called an "Aston" after the most dominant manufacturer there, whereas in the United States "Chyron" is practically synonymous with a CG for the same reason. (See also SUPERIMPOSITION.)

charge-coupled devices (CCD)   devices that change light into electronic signals. They consist of one to three chips, which break down the television picture into thousands of pixels (picture elements). The horizontal and vertical photosensitive elements cross within the tiny device and create the image. The number of pixels determines the definition and quality of the picture. The CCD chips cost less than the conventional high-quality pickup tubes and were initially used in consumer home video CAMCORDERS. As their definition and resolution improved, CCDs became popular in professional broadcast operations where three-chip CCD cameras are now the norm.

chargeback system a method of accounting to calculate the funds used in producing media for other departments within a company or institution. The basic overhead for the media center is covered by the parent institution, but the center charges a service fee to any department or division borrowing videocassettes or requesting graphics or television production and an interdepartmental transfer of funds is made.

checkerboarding a program scheduling strategy whereby individual programs from a series are transmitted at the same time on alternate days of the week. The technique differs from STRIPPING programs or running them ACROSS THE BOARD each day of the week.

cherry picking a program acquisition practice that involves the selection of programs from a variety of sources. The objective is to pick and choose the best programs for scheduling on a cable system or television station. The technique is used by INDEPENDENT STATIONS that are free to choose from a number of distributors in acquiring the programs that make up their broadcast schedules.

Chicago school (of journalism) an exacting method of journalism promulgated by the legendary City News Bureau of Chicago. The bureau acts as a boot camp for aspiring journalists. No reporter ever turns in a story without specifying the exact age, address, and middle initial of the individual in the tale. The standard for accuracy was exemplified by a long-time night editor of the City News Bureau who was said to bellow at his quaking charges, "Your mother says she loves you? Check it out!"

Chicago school (of television programming) programs in the 1950s that were known for their imaginative style and for the creative ingenuity used in developing them in relatively primitive circumstances. All of the stations and network divisions in Chicago in the early days of television contributed to the city's reputation as a center of new production techniques at that time.

Child Protection Restoration and Penalties Enhancement Act a law that requires ADULT VIDEO producers and wholesalers of X-rated movies to provide proof that the performers in the films are at least 18 years of age. They must label the boxes containing prerecorded videocassettes accordingly.

Children's Television Act of 1990 a law that limits the amount of advertising time for shows produced primarily for children under the age of 12. The Act also conditioned the renewal of broadcast licenses on the extent to which the broadcaster has served "the educational and informational needs of children," directed the FCC to study the role of "program-length commercials" on children, and established an endowment fund for children's programming.

Children's Television Workshop (CTW) a private, nonprofit organization that produces programs for children. The programs combine information and instruction with entertainment. "Sesame Street," which premiered on PUBLIC TELEVISION (PTV) stations in 1969 with the Muppets as the centerpiece, is the quintessential CTW show.

churn   the rate of turnover of customers who do not renew their subscriptions to a cable TV operation, MULTICHANNEL MULTIPOINT DISTRIBUTION SERVICE (MMDS) system, or LOW POWER TELEVISION (LPTV) STATION that offer a scrambled signal. The churn is measured for monthly or yearly periods and gauges the effectiveness of the PAY-TV company. The churn rate is expressed as the percentage of subscribers who request that the company discontinue its service. (See also SCRAMBLING.)

circulation   See CUME.

clapstick   a small chalkboard with a hinged top used in film productions. The title of the program and scene number and the take are written on it, along with other information that identifies the scene being shot. Before the scene is undertaken, a production assistant claps the hinged top against the base of the board as it is shot by the camera. Because the sound is usually recorded separately in film production, the sharp noise of the clap of the board is later used to synchronize the picture and sound. In television productions, a clapstick or poster board card is used to visually identify

the scene but the clap sound is not needed.

Clear–TV   a nonprofit group that organizes product boycotts of the sponsors of programs that contain "gratuitous sex/violence and anti-Christian stereotyping." Officially known as the Christian Leadership for Responsible Television (Clear–TV), this Wheaton, Illinois-based organization works to persuade the networks to eliminate "antifamily" programs.

clearance   See STATION LINEUP.

client   a term used synonymously with ACCOUNT in the advertising industry. An ADVERTISING AGENCY or a television or cable operation has clients or accounts but never customers or patrons, because the term "client" serves to add more dignity and prestige to the relationship.

clip   See FILM CLIP.

clipping   an unethical and illegal practice of cutting away from a program transmitted by a network to insert and transmit local COMMERCIALS. Clipping usually occurs at the end of a program while the credits are being shown.

closed captioning   the process of encoding written words into a television or video program for display during the viewing of the program. The captions are printed lines KEYED on the bottom of the screen to explain the plot or to condense the dialogue for the benefit of the hearing impaired. The captions cannot be seen by the viewer without a DESCRAMBLER DECODER, which is attached to the television set (hence the term "closed").

Closed captioning uses the vertical BLANKING interval in the television SCANNING LINE process to transmit and display the writing.

After July 1, 1993, all sets sold in the United States with a 13-inch or larger screen must have an internal decoder and the capability of displaying closed captions. (See also NATIONAL CAPTIONING INSTITUTE [NCI].)

closed-circuit television (CCTV) a type of private cable service that transmits and receives a signal in a closed loop. The signal is sent through a COAXIAL CABLE that connects one or many different locations to the origination point.

All cable systems are theoretically closed-circuit systems but the term is commonly applied to operations in a small geographic area that are used for a particular purpose. CCTV systems are often a part of a CORPORATE TELEVISION or AUDIOVISUAL COMMUNICATIONS operation. For example, they can be installed within a manufacturing plant or a school system. Universities often operate CCTV systems between classrooms on a campus.

closeup (CU) a television or film shot in which the subject dominates the screen. Sometimes called a tight shot, a CU often shows only the

head of the person in the picture. An even closer view, called an extreme closeup (ECU) with less space around the subject such as a shot of the eyes, is sometimes used for dramatic effect. (See also COMBINATION SHOT, FRAMING, LONG SHOT [LS], and MEDIUM SHOT [MS].)

cluster analysis a method of statistical geodemographic research and analysis, often based on ZIP codes. In this study, computer programs group people by common characteristics on the assumption that people of similar backgrounds will live near one another and have similar tastes, income levels, behavior, and purchasing patterns. (See also ACORN and PRIZM.)

clutter the practice of scheduling a large number of COMMERCIALS, PSAS, and ADVERTORIALS along with PROMOS and STATION IDENTIFICATIONS (ID) within a COMMERCIAL TIME period. Each unit competes for the viewer's attention and within the mass of information, the impact of any single message is usually lost in the clutter.

CLV (constant linear velocity) See LASER VIDEODISC (LV).

coaxial cable the cable in cable systems, commonly called "coax" (pronounced co-ax). This type of flexible connecting cable is the backbone of BROADBAND COMMUNICATION (SYSTEMS). Ranging in size from a quarter-inch to one-inch in diameter, the cable is composed of a central solid conductor surrounded by a hollow cylinder. Both are encased in a plastic outer shell. The energy travels between the inner two conductors and is shielded, thereby reducing energy loss.

Small coax cables are also used in many other aspects of television to connect cameras, video recording equipment, MONITORS, and other electronic gear. Television studios and control rooms are a maze of coaxial cable. (See also ATTENUATION, CABLE DROP, FEEDER CABLES, and TRUNK LINES.)

color bars　vertical bars of different colors used to test and balance color in television cameras, videotape machines, and other production gear. They consist of seven (sometimes eight) bars of pure white, yellow, blue-green, green, reddish purple, red, and blue (and sometimes black). They can be created electronically or by focusing the camera at a cardboard test or color chip poster board containing the bars. (See also GRAY SCALE.)

color temperature　the ratio between the six main colors (ranging from red to blue) in the color spectrum of a light source for a television picture. Color temperature is measured by degrees KELVIN (K), which range from 2,800x K to 7,000x K. Daylight at midday measures about 5,600x K, which is cooler (more blue) than the 3,200x K (more red) that is usually used as the reference for color cameras under the artificial light from SCOOP LIGHTS and SPOTLIGHTS in a television studio.

Columbia Broadcasting System (CBS)　See CBS INC.

Columbia Pictures Entertainment (CPE)　a mammoth entertainment corporation dating from 1919. By 1924 the company had grown into Columbia Pictures Corporation and prospered in Hollywood throughout the 1930s and 1940s. Anticipating the growth of television, the firm formed a subsidiary, Screen Gems, in 1951 for the purpose of producing television programs and distributing them (along with Columbia films) to the new medium. During the next decade the company began to develop what is now one of the largest television libraries. The Screen Gems name was dropped in 1976, and in 1982 the Coca Cola company purchased Columbia. In 1987 Columbia joined HOME BOX OFFICE (HBO) to form the first new major motion picture studio in decades, Tri-Star Pictures. Later in 1987 a division of the Coca Cola company and Tri-Star Pictures formed Columbia Pictures Entertainment (CPE), which now operates the Tri-Star and Columbia Pictures Studios. In November 1989 CPE was purchased by Sony USA Inc. (a subsidiary of the SONY CORPORATION), and two years later the new owner renamed the venerable Hollywood institution Sony Pictures Entertainment.

combination shot　a television and film camera shot that is a combination of a LONG SHOT (LS) and a CLOSEUP (CU). Called a "combo" for short, it is usually used in dramatic or musical-variety shows where one person is in the foreground while another or others are seen behind the main action. (See also FRAMING.)

Comcast One of the cable industry's top MULTIPLE SYSTEM OPERATORS (MSO). The company owns cable systems clustered in the Northwest, the Southeast, the Midwest, and in California.

Comedy Central a BASIC CABLE network that was originally named Comedy TV. The network was formed by the 1991 merger of the Comedy Channel, which was owned by HOME BOX OFFICE (HBO) and its parent TIME WARNER, INC. and the comedy network HA!, owned by VIACOM. The new channel began with a mixture of programming from the libraries of both of the original operations, but now also provides original programming.

commercial a brief advertising message on a cable or television operation. A commercial combines motion, sight, sound and words and is designed to persuade or entice the viewer to take a particular action or to purchase the goods or services of a company. In the United States, television commercials are scheduled within and between programs and are either 10, 15, 30, 45, 60, or 90 seconds in length. They are scheduled throughout the day and evening hours and are often called "SPOT (ANNOUNCEMENTS)" or "SPOTS." (See also ADVERTORIAL, COMMERCIAL TIME and PUBLIC SERVICE ANNOUNCEMENT (PSA).

commercial protection the practice of segregating COMMERCIALS for similar products or services of different companies. Either as a requirement of a SPOT (CONTRACT) or through standard industry practices, SPOT (ANNOUNCEMENTS) for competing products on cable or television operations are scheduled at least 10 minutes apart from one another.

commercial time time periods that are set aside for COMMERCIALS on a television or cable operation. The advertising intervals are scheduled before, after, or within programs from SIGN-ON to SIGN-OFF. Commercial time is an integral part of the STATION BREAKS on commercial television stations.

Commission on Instructional Technology (CIT) a nine-member commission of the U.S. government that was directed to "recommend to the president and Congress specific actions to provide for the most effective possible application of technology to American education." The group's report, which was submitted in 1970, made many worthwhile recommendations but few of them were acted upon. (See also PUBLIC BROADCASTING ACT.)

commissionable services provided by an ADVERTISING AGENCY for which it receives a commission. Commissionable income differs from the fees charged by agencies to their clients for research or production.

Agencies that are recognized as legitimate by television and cable operations are given a percentage (usually 15 percent) discount off the regular RATE CARD charges when they buy COMMERCIAL TIME for their clients. In such instances the regular rates of the television and cable operations are said to be commissionable. (See also ADVERTISING AGENCY COMMISSION.)

Committee on Local Television Audience Measurement (COLTAM) a committee of the NATIONAL ASSOCIATION OF BROADCASTERS (NAB) that was formed to investigate better methods of audience research at the local station level. After a three-year study financed by the industry, the committee recommended a new written DIARY SYSTEM in 1990 to replace the systems then used by A. C. NIELSEN and the ARBITRON COMPANY.

common carrier the types of communications services licensed by the FEDERAL COMMUNICATIONS COMMISSION (FCC) or regulated by state public utility agencies that do not maintain editorial control over the content of any transmissions. Common carrier companies offer communications services to the public at published tariffs (rates), which are regulated by the FCC or the state authority.

The most familiar common carriers are the telephone, telegraph, and SATELLITE services.

Commonwealth Broadcasting Association (CBA) a nonprofit organization that consists of national broadcasting organizations in 50 British Commonwealth countries and territories.

Communication Association of the Pacific See WORLD COMMUNICATION ASSOCIATION (WCA).

Communications Act of 1934 the law that established the FEDERAL COMMUNICATIONS COMMISSION (FCC) and gave it authority to regulate broadcasting and other communications industries, including COMMON CARRIERS such as telegraph and telephone services.

The primary stated purpose of the Act was "to provide for the regulation of interstate and foreign commerce in communication by wire and radio so as to make available (so far as possible) to all the people of the United States, a rapid efficient nationwide and worldwide wire and radio communication service with adequate facilities at reasonable charges."

One part of the Act defined "radio communication" as the "transmission by radio of writing, signs, signals, *pictures* [emphasis added], and sounds of all kinds . . ." thereby allowing for the legislation to cover television many years later. The most quoted aspect of the Act requires broadcast stations to operate in the "public convenience, interest, or necessity."

The Act gives the FCC authority to allocate various parts of the electromagnetic spectrum and to assign FREQUENCIES for various services. The Commission is also authorized to license individuals, companies, and organizations to operate broadcast stations and to establish criteria for the granting of a license. The legislation, however, specifically prohibits the FCC from interfering with free speech or engaging in the censorship of programs.

There have been many revisions and amendments to the Act since its passage in 1934 leading to the often-used Washington phrase, "the Communications Act of 1934 as amended." In 1977 a House Committee called for a massive comprehensive rewrite of the Act. It continues, however, to be the arbiter of the nation's policy for communications. (See also FCC LICENSE, LICENSE RENEWAL, and SPONSORSHIP IDENTIFICATION RULES.)

Communications Satellite Corporation (Comsat) a private, profit-making company, formed in 1962 as the first SATELLITE firm to be chartered by the U.S. government. Comsat represents the nation in a consortium of 119 countries that finances a system of communications satellites. The consortium, called the International Television Satellite Organization (Intelsat), cooperatively finances the launching of communications satellites.

community antenna relay system (CARS) See MICROWAVE RELAY.

community antenna television (CATV) a phrase used to describe the first cable television systems in the late 1940s. They were local ventures created to bring in better reception of television pictures. A master antenna was raised on a mountain peak and the houses in the valleys were linked to it by COAXIAL CABLE, enabling home viewers to receive a clear picture. Thus the system was labeled "community antenna television." The term was replaced by "cable television" in the late 1950s but it is still occasionally used in the industry. (See also CABLE TELEVISION [DEFINITION] and PAY PER VIEW [PPV].)

Community Antenna Television Association (CATA) a nonprofit TRADE ASSOCIATION, CATA represents many of the smaller independent cable systems in the United States. Acting as a counter organization to the NATIONAL CABLE TELEVISION ASSOCIATION (NCTA), CATA represents its members' interests before Congress and the FEDERAL COMMUNICATIONS COMMISSION (FCC).

Community Broadcasters Association (CBA) a nonprofit organization formed in 1988 by professionals in the nascent LOW POWER TELEVISION (LPTV) industry. Its objective is to encourage the growth of LPTV stations and to promote the industry.

community broadcasting See LOW POWER TELEVISION (LPTV) STATIONS.

community service grants (CSG) funds given annually to each qualified PUBLIC TELEVISION (PTV) and radio station by the CORPORATION FOR PUBLIC BROADCASTING (CPB) to enhance the local station's ability to serve its community. They constitute an indirect form of federal funding for national programming because the stations normally use a large portion of their annual grants to purchase programming through the PBS PROGRAM FUND.

community station a type of PUBLIC TELEVISION (PTV) station that is licensed to a nonprofit organization in a community by the FEDERAL COMMUNICATIONS COMMISSION (FCC). A community station is owned and operated by an organization that was established for that specific purpose.

compact disc (CD) a consumer audio product, developed by the SONY CORPORATION and Philips in 1933, which revolutionized stereo sound. Using a DIGITAL COMMUNICATIONS process, CD recordings are near-exact replicas of the original audio performance. The recordings are made on magnetic tape and transferred to a small round disc of either three or five-and-a-quarter inches in diameter. The digital information is embedded in small pits on the disc and is read by a LASER while the disc rotates rapidly in a counterclockwise direction. Over the years the audio capability of the CD has been enlarged to include visual and textual information.

comparative hearings formal hearings held at the FEDERAL COMMUNICATIONS COMMISSION (FCC) when there are mutually exclusive applicants for CONSTRUCTION PERMITS (CP) to build a broadcast station or when station licensees are challenged at LICENSE RENEWAL times. The hearings are presided over by an administrative law judge (ALJ) who receives documents, affidavits, and testimony

from the competing parties. This adjudicating process seeks to determine which of the applications would best serve the public interest.

component video system/recording a method of processing and recording and playing back a video signal. The technique involves the separate recording of the chrominance (color) and luminance (black-and-white brightness) aspects of a television signal. The color video signals are recorded separately from the black-and-white signals on two separate tracks prior to combining them. This reduces the loss of detail in the image and the result is a much superior picture in terms of sharpness and resolution. The method is now preferred over the older COMPOSITE VIDEO signal process.

composite video system/recording an initial method of handling television signals that is used in processing a video signal, recording it, and playing it back. In black-and-white television, composite video consists of the picture signal as well as the synchronization (SYNC or timing) and BLANKING pulses all combined into one signal. In color television, additional color picture and synchronizing information is added. A SYNC GENERATOR is used to generate the single composite signal from the noncomposite video pulses. The combined signal can then be decoded and processed in various pieces of equipment. The signals must then be reencoded, however, to be usable and this constant processing distorts and degrades the signal.

compulsory license system See COPYRIGHT ACT OF 1976.

computer-assisted instruction (CAI) a term from the 1960s that described the use of computers for individual instruction in a formal academic environment such as a school or college. It referred to experimental and innovative programs using mainframe computers. The operations were cumbersome and expensive and CAI languished until the advent of the personal computer (PC).

Some professionals in the computer field today prefer to use the term "computer-assisted learning" or "computer-based education." In the business world of CORPORATE TELEVISION, CAI is often called "computer-based training (CBT)" but the principles and processes are the same.

Comsat See COMMUNICATIONS SATELLITE CORPORATION (COMSAT).

condenser microphone the most common microphone used in television production. The condenser mike is often built into portable cameras and is also used as a lapel and tie-clip mike for newscasters and interviewees. It is sometimes called an electrostatic or capacitor mike.

Consolidated Metropolitan Statistical Area (CMSA) a designation of a geopolitical area, determined by the Office of Management and Budget (OMB) of the U.S. government. A CMSA consists of two or more contiguous METROPOLITAN STATISTICAL AREAS (MSA). The designation is used as a basis for similar designations by audience-reporting services such as the ARBITRON COMPANY and A. C. NIELSEN. (See also MARKET.)

Consortium of College and University Media Centers (CCUMC) a nonprofit organization of AUDIOVISUAL COMMUNICATIONS departments or divisions at institutions of higher education that seeks to encourage the use of film, video, and other materials in education.

construction permit (CP) a written authorization from the FEDERAL COMMUNICATIONS COMMISSION (FCC) that allows the applicant for a broadcast license to build a full-power radio or television station, a LOW POWER TELEVISION (LPTV) STATION or a MULTICHANNEL MULTIPOINT DISTRIBUTION SERVICE (MMDS) system. The applicant is given a specific period of time to actively build and test the electronic transmission of the station. After construction and a PROOF OF PERFORMANCE is completed, the operators apply for and receive an FCC LICENSE to operate the station.

consumer electronics a broad term encompassing the personal electronics products that bring education, information and entertainment to the public. These products include video devices such as television sets, videocassette and videodisc machines, CAMCORDERS, and TELEVISION RECEIVE ONLY (TVRO) dishes as well as audio components, radios, COMPACT DISC (CD) players, tape recorders and home information equipment such as personal computers, photocopy and fax machines, and telephone answering devices. In addition the field includes calculators, electronic games, and electronic musical instruments.

Consumer News and Business Channel (CNBC) a BASIC CABLE SERVICE network, operated by NBC, that concentrates on the world of business and personal finances. Its programming includes stock market reports, interviews and discussions.

continuity   words and images used between programs on a cable system or television station. The information promotes upcoming programs and provides STATION IDENTIFICATION (ID) information during STATION BREAKS. The objective of continuity is to provide a seamless transition between programs with a mixture of IDs, PROMOS and COMMERCIALS.

Conus Communications   a SATELLITE NEWS GATHERING (SNG) cooperative of local television stations and news organizations operating as a profit-oriented business. It specializes in live remote news coverage from anywhere in the world. CONUS (satellite-engineering language for "**continental U.S.**") provides a Washington D.C. news feed and breaking news cut-ins, compiled in part from reports from member stations via KU-BAND SATELLITES.

converter   a device used in a cable system to change the FREQUENCIES of an electronic signal. At the HEADEND of a cable system, the device converts the television signals

that are received from a SATELLITE or MICROWAVE RELAY station into other signals that can in turn be sent to the system's subscribers. At the receiving end of the cable operation, the viewer's home, a converter is used to enable reception of a wide variety of programs.

A small rectangular box-like device is used to convert those cable channels to channels that can be received on the subscriber's TV set. The set is tuned to a vacant channel (such as channel 3) and all of the channel switching is then done from and by the converter. The controller for the converter contains a series of push buttons and the viewer selects the desired cable channel by pressing the appropriate button.

Some of the newer converters can also be used to turn the television set off and on and to control the audio volume. Many of the newer devices are ADDRESSABLE CONVERTERS and can be controlled from the headend of the cable system.

cookie   a pattern cut out of metal or wood that is placed in front of a light source, thus projecting the pattern on the wall or backdrop of a television studio. Formally known as a "cucoloris" (pronounced kook-a-LOR-iss), the small three-inch (in diameter) pattern is usually placed in a special SPOTLIGHT designed to be used with the device.

cool light   a lighting instrument that provides high light output with less heat and permits film to consistently capture more realistic color. In addition to providing cooler working conditions, the lights emit a steady stream of pale blue light at about 5,600x KELVIN. (See also HMI LIGHT.)

cooperative (co-op) advertising the practice in which the manufacturer and the retailer share the cost of advertising a product or a service. A national firm such as a major film studio or PROGRAM SUPPLIER in home video prepares an advertisement for a movie and the video store adds its name and places the ad in the local newspaper. The store is reimbursed for a portion of the cost of the media placement by the larger firm.

Co-op advertising is also used by the networks and SYNDICATORS to help build the audience for the programs they distribute. Print ads are placed or SPOT (ANNOUNCEMENTS) aired by the local station or cable system and a portion of the cost is billed to and paid for by the distributor of the program. (See also DOUBLE BILLING, MINIMUM ADVERTISED PRICE [MAP], TUNE-IN ADVERTISING, and WHOLESALER.)

coproduction the sharing of costs of the production of programming between two or more stations, networks, or syndicators.

copy Creative text or dialogue designed to persuade a consumer to buy something or to promote a program, cause, or product. Copy can take the form of a STATION IDENTIFICATION (ID), a PROMO or a COMMERCIAL. Its purpose is to communicate ideas and concepts in an original way, usually in short concise sentences or phrases. The term was orig-

inally used in the newspaper industry when stories were written in longhand by reporters and had to be transcribed into type (copied) by linotype operators in order to be printed.

copyright principles and laws that protect a creator's property from duplication, use, or sale by someone without the permission of the owner. Under U.S. law, a work is deemed to be copyrighted upon its creation. It need not be published, broadcast, or distributed in order to be considered copyrighted. Full protection, however, can be achieved by the registration of the work with the Copyright Office of the Library of Congress.

Copyrights protect only the expression of an idea, not the idea itself. They can cover art, motion pictures, television programs, books, music, photographs, commercials, and advertisements along with other intellectual properties. Titles or names, however, cannot be copyrighted. They are protected by the Lanham Trademark Act of 1946. If a copyright on a work expires, the work passes into the PUBLIC DOMAIN.

Copyright Act of 1976 a legislative act seeking to balance the rights of the creators of material with the rights of the public to have access to the material.

When this legislation became effective in January 1978, it replaced the 1909 COPYRIGHT law that had protected intellectual property in the United States for many years. The new law took into consideration some of the new electronic media, including cable television, and established a national standard for dealing with works that were protected by copyright. The Act maintained the copyright life of older works at 28 years from first use, but extended the length of the renewal period to 47 years. For works created after 1978, the copyright protection is for the duration of the creator's life plus 50 years. (See also OFF-AIR VIDEO RECORDING, PUBLIC DOMAIN, and PUBLIC PERFORMANCE OF COPYRIGHTED VIDEO MATERIAL.

Copyright Royalty Tribunal (CRT) a government agency established by Congress in 1978 as part of the COPYRIGHT ACT OF 1976 to collect copyright fees from cable systems and distribute the fees to copyright holders. The CRT sets rates and determines the percentage of the total fees that will be distributed to the different types of copyright holders.

corporate television the use of television in business, alternately called "corporate television," "corporate video," "private television," "organizational television," or "business television." When the medium was first used in the business world it was more simply labeled "industrial television" or "nonbroadcast television." Under whatever label, communications technology in business has grown consistently in the past two decades.

corporate video See CORPORATE TELEVISION.

Corporation for Public Broadcasting (CPB) a private, nonprofit corporation created by Congress in the PUBLIC BROADCASTING ACT of 1967 to facilitate the growth of PUBLIC TELEVISION (PTV) and radio in the United States and to insulate such broadcasting from external political interference and control. The CPB is not a federal agency but it receives its budget from congressional appropriations.

cosponsor See PARTICIPATING SPONSOR.

cost per rating point (CPRP) an audience research method that helps assess the effectiveness of COMMERCIALS. The system measures the efficiency of a SPOT by comparing the RATINGS generated during the period the spot was aired with the cost of the commercial time. Often shortened to cost per point (CPP) in everyday use, the CPRP is obtained by dividing the dollar cost of the spot by the GROSS RATING POINTS (GRP) achieved, and is the cost of reaching one percent of the audience one time. (See also COST PER THOUSAND [CPM].)

cost per thousand (CPM) the cost (to an advertiser or advertising agency) of reaching 1,000 homes or people. The C(cost P(per) M(Roman 1000) measures the efficiency of an advertising message. The CPM is derived by dividing the cost of the commercial time by the audience in thousands. If, for example, the cost of a commercial is $75,000 and the audience seeing it was 5,200 in thousands (5,200,000 people divided by 1,000), then the CPM is $14.42. (See also M.)

couch potatoes individuals who plop themselves down in front of a television set for marathon viewing sessions. Armed with snacks for sustenance and a remote control, they often spend weekends GRAZING and ZAPPING from channel to channel. Although the phrase can be applied to either sex, it is most often used by wives to describe their husbands' behavior.

Council on International Nontheatrical Events (CINE) a nonprofit organization that coordinates the selection and submission of short American DOCUMENTARY, NONTHEATRICAL FILMS and videotapes to some 80 international film events and festivals. CINE sponsors two competitions each year in the spring and fall and bestows the Golden Eagle awards.

counterprogramming a program scheduling technique by which programs appealing to completely different audiences and DEMOGRAPHIC groups are scheduled directly opposite those appearing on competing channels. The strategy is often used by stations or systems that have fewer viewers and is designed to attract segments of the audience that are dissatisfied with the programs in that DAYPART on the competing channels.

courseware software used in a TELECOURSE, including print materials such as manuals, study guides, textbooks, tests, and workbooks. Courseware often comprises the bulk of the telecourse.

coverage area the geographic area in which the signals from a television station can be received through the air by an audience. Because television signals are line-of-sight waves, they reach a point where reception is hazy and eventually impossible. Most television signals can be picked up within 20 miles of the station with an indoor receiving antenna. At 30 miles, a high outdoor antenna is necessary. Most stations cannot be seen 60 miles from a transmitting tower, and thus the signal is lost at that distance.

The FEDERAL COMMUNICATIONS COMMISSION (FCC) regulates the extent and pattern of the coverage area to ensure that the broadcasts of stations on the same or nearby FREQUENCIES do not interfere with one another. The FCC classifies the coverage area of all television stations in terms of Grade A and Grade B contours. Grade A contours consist of geographic areas in which a satisfactory signal can be received 90 percent of the time in at least 70 percent of the receiving locations within the contour. Grade B contours encompass an area where reception is adequate 90 percent of the time in at least 50 percent of the locations.

CPB-qualified station a local PUBLIC TELEVISION (PTV) or radio station that is eligible for an annual COMMUNITY SERVICE GRANT (CSG) from the

CORPORATION FOR PUBLIC BROAD-
CASTING (CPB). To fulfill the requisite
conditions, a station must meet certain standards of operation.

cradle head   a sturdy camera support that sits atop a TRIPOD or STUDIO
PEDESTAL camera mount and supports the base of the camera, which
in turn sits atop its cradle-like shape.
The cradle head provides for the
smooth operation of the camera,
particularly for TILT shots. (See also
FLUID HEAD and FRICTION HEAD.)

cradle-to-grave   a   programming
term indicating that the audience for
a show is universal. Children, adults,
teenagers, parents, and all DEMO-
GRAPHICS and PSYCHOGRAPHICS are
represented in the viewing audience.
The World Series is a cradle-to-
grave program.

crane   a type of camera mount. It
is a large metal device fitted with
three or four wheels and a boom
arm that allows a camera to be raised
smoothly from near-floor level to 10
feet in the air.

crawl   a graphic effect that moves
vertically or horizontally over the
screen. Crawls are usually used to
list the CREDITS for a production at
the end of a program. Names of the
cast and crew are SUPERIMPOSED or
KEYED over the final shots and
"crawl" up or down.

Crawls are also electronically generated to move horizontally across
the bottom one-eighth of the screen,
usually from left to right. They are

used for special news flashes, to announce delays in the program schedule or to provide other viewer
information.

credits   the list of names and titles
of the people who worked on or
contributed to a production. Credits
are usually presented in the form of
a visual CRAWL or series of graphics
at the end of a program and are
sometimes accompanied by a VOICE-
OVER reading the names of the cast.

crop   to trim the information near
the edges of the screen in television
production. The term was borrowed
from still photography. A camera operator will be directed to crop a
graphic at the top or bottom or the
sides by moving in to lop off a small
portion of it. (See also FRAMING.)

cross fade   the slow transition between one production element and
another. It is most commonly used
in audio where one sound decreases
while another increases and gradually replaces the initial sound. In
lighting, one source is dimmed while
another slowly becomes brighter and
replaces the first. The same effect in
video is called a DISSOLVE. (See also
CUT, FADE IN/FADE OUT, and SEGUE.)

cross-media buy   the purchase of
advertising time and space in more
than one medium. A CAMPAIGN can
be devised to purchase SPOTS on cable and broadcast television and
space in newspapers and magazines.
The term usually implies the use of
two different communications media

(for example, newspapers and television), while multimedia buys involve more than two media elements such as newspapers, magazines, television, and radio.

cross-ownership rules FEDERAL COMMUNICATIONS COMMISSION (FCC) rules issued in 1975 prohibiting the ownership of a daily newspaper and broadcast operation in the same community. The intention was to expand on the DUOPOLY RULE to promote the diversification of all media.

The rule resulted in a sharp drop in local newspaper ownership of television stations. No similar FCC restrictions apply to any newspaper-cable system cross ownership but cable systems may not be owned by television stations or telephone companies in their service areas.

crosstalk the unwanted transfer of a signal transmitted on one channel or circuit to another channel or circuit. The undesired transfer creates BACKGROUND NOISE and interferes on the second channel.

cue audio and visual signals that trigger all aspects of a television production. Cues help coordinate the various elements in a production by preparing the talent and crew and by calling for action and execution on their part. Some cues are given verbally, others by hand signals. They direct announcers or talent to speed up or slow down. Actors give cues to one another in the form of lines in a drama. Audio and video tape recordings are "cued up" (ready to be inserted into a program) and cues

are given to on-camera interviewers to stretch the conversation out or wind it up and conclude the show.

cue cards large pieces of white cardboard containing the lines to be spoken or lyrics to be sung by actors or performers in a television production. The words are printed with a felt-tip marker in large clear letters and the cards are held by a production assistant near the camera lens. They are sometimes known as "idiot boards."

cume an abbreviation of "cumulative audience." It is the measurement of a television audience over a specified period of time. It is also known as the "net audience" or "unduplicated audience."

The cume constitutes the "reach" or "circulation" of the station, and represents the number of people reached at least once during the measurement period.

cumulative audience See CUME.

CUPU leased access channels channels designated for "commercial use by persons unaffiliated" with the cable operator. One of the major innovations of the CABLE COMMUNICATIONS POLICY ACT OF 1984, the concept is designed to promote program diversity within a cable system. Section 612 of that Act requires cable operators with 36 or more activated channels to designate some for commercial CUPU purposes. The system operator can have no editorial control over the CUPU channels.

The intent is to open up some channels for use by programmers who have been denied access to a cable operation because of the content of their programs. The system must provide them with the channels at an appropriate price and under reasonable terms and conditions. The CUPU idea of commercial access to cable distribution complements the noncommercial PEG CHANNEL requirements in providing for increased diversity in cable programming.

curriculum materials   See AUDIO-VISUAL COMMUNICATIONS.

cut   an instantaneous change from one picture to another. It signifies an abrupt change in action or pace. The term is derived from film editing techniques in which the quick alteration in shots is accomplished by physically cutting the film and splicing the two different shots together.

In television production the technique is sometimes called a "take" because the command to execute the action by punching a button on a SWITCHER is often "Take one!" (for camera one) or "Take two!" (for camera two). The word cut is also used in television and film production as a command by a director to immediately stop the action on a set.

cutaways   shots that are used in television and film productions to avoid JUMP CUTS and to make the editing process easier and the transitions from shot to shot more graceful and smooth. A LONG SHOT is often inserted between two CLOSEUPS (CU), to cut away from the intense action for a moment or to hide an error in continuity between the two closeups. Cutaways are most often used in one-camera taped interviews where shots of the reporter nodding or asking questions are shot after the interview is over and later edited into the program to help disguise audio edits.

cyclorama   a staging piece used in television, stage, and film production, called a "cyc" (pronounced "sike") for short. It is a continuous floor-to-ceiling background made of cloth or plasterboard that creates an illusion of infinity by eliminating a visual frame of reference. A cyc surrounds the staging area in a studio in the rear and on one or two sides without visible corners and seemingly also melts into the floor.

D

dailies   an assemblage of footage that has been shot on any given day in film production. The dailies are projected for the director, perform-ers, and crew and are used to review the style, technique, and quality of the production and the performance of the actors.

daily topicals the short bits of news items that tease the audience about the stories scheduled for presentation on that evening's newscast. During STATION BREAKS, one or two announcers appear on camera, often from the newsroom, to give one-sentence summaries of news stories that will be explored in more depth later.

DAT (digital audio tape) technology that was introduced in the United States by the SONY CORPORATION in 1990. The format employs the same DIGITAL COMMUNICATIONS techniques used in video recording technology. Sound is stored in a series of numbers and the result is superior, pristine audio with much less distortion than is possible with ANALOG COMMUNICATIONS techniques. DAT units use a very small tape cassette, about two inches by two-and-three-quarter inches.

Data Discman a small, handheld, electronic device that resembles a palm-size computer with a miniature keyboard. It is essentially a paperless book. Books, in the form of three-and-a-half-inch (in diameter) optical disks, are inserted into the machine, which displays the pages on a screen about the size of a business card. One disc can contain 100,000 pages of text. Pages are turned or recalled in response to commands typed on the keyboard.

daypart a daytime period of programming in a television or cable operation. Although the specific determination of the times and their descriptive labels vary among stations, agencies, and RATINGS companies, most agree that the programming day is divided into seven or eight dayparts. The periods most often cited in the industry are commonly called DAYTIME, EARLY FRINGE, KID FRINGE, LATE FRINGE, LATE NIGHT, PRIME TIME, and PRIME-TIME ACESS.

daytime a program scheduling period on television stations, networks, and cable systems that is usually recognized as being between 9:00 A.M. and 4:30 P.M. eastern standard time (EST) Monday through Friday.

dealer imprint the name and address of a local video store that is imprinted on a sell sheet, a national advertisement, or a pamphlet. The promotional piece is provided by a manufacturer of electronic equipment or a PROGRAM SUPPLIER, and the local store puts its name on it with a stamp or a stick-on tag.

decibel (dB) a measure of the power of one electronic signal compared to another. It is the logarithmic unit that expresses the signal-strength ratio between the two. A bel (named after Alexander Graham Bell) is equal to 10 decibels, a more useful way of measuring signals. A decibel, then, is defined as 10 times the logarithm of the ratio of the two powers. Expressed in algebraic terms, it is $dB = 10 \log_{10}(P_1/P_2)$. The higher the resulting number, the greater the signal strength.

decoder See DESCRAMBLER/DECODER.

dedicated channel one or more unused channels on a cable system reserved for later use.

delayed broadcast a program transmitted on a television station or cable system at a later time than it actually occurs or is scheduled. The program is recorded on videotape for subsequent broadcast.

demo reel sometimes called a sample reel or simply a "demo." This collection of COMMERCIALS, programs, or segments of programs is compiled by an actor, director, producer, ADVERTISING AGENCY, or INDEPENDENT PRODUCTION COMPANY to be shown to prospective CLIENTS. "Demo" is short for demonstration and "reel" stems from the fact that such compilations were originally made on a 16mm reel of film. Today demos are more often recorded and presented on videocassette, but the film term is still used.

demographics the classification of an audience by socioeconomic characteristics such as age, sex, occupation, income, race, and family size. The audience segments are defined in different ways by research companies such as A. C. NIELSEN and the ARBITRON COMPANY. They are extremely important to advertisers and ADVERTISING AGENCIES. The term is often shortened to "demos." (See also CLUSTER ANALYSTS and PSYCHOGRAPHICS.)

depth of field the distance from the nearest object in a picture to the farthest object that is in sharp focus.

The degree to which this area surrounding the main subject is in focus denotes the extent of the depth of field. It is determined by the distance from the camera lens to the subject, the FOCAL LENGTH of the lens, and the F-STOP that is being used.

descrambler/decoder devices for transforming unintelligible, fragmented, or coded pictures into cohesive images in both over-the-air and cable technology. Broadband or RF descramblers are used to reconstruct the cable signal in the subscriber's home. The descrambler can be a small rectangular box that sits on top of the TV set or an internal part of the CONVERTER unit. The descrambler either reconstructs the signal directly or modulates and remodulates the signal to give the picture coherence and an intelligible form. (See also MODULATION.)

Descriptive Video Service (DVS) an aspect of the SEPARATE AUDIO PROGRAM (SAP) service that provides ongoing simultaneous narration of on-screen television action for the blind. The project uses the separate audio channel to offer descriptions of the setting and action for selected dramas.

designated market area (DMA) counties that surround a metropolitan center, as defined by the A. C. NIELSEN COMPANY. More than 200 such geopolitical areas are so designated in the United States. The major viewing audience for the stations in that particular area are within the specific DMA. DMAs are similar to the AREAS OF DOMINANT INFLUENCE (ADI) of the ARBITRON COMPANY. Both correspond to the METROPOLITAN STATISTICAL AREAS (MSA) and the CONSOLIDATED METROPOLITAN STATISTICAL AREAS (CMSA) as defined by the Office of Management and Budget of the federal government.

desktop video   the equipment and the process used for the production of smaller-format nonbroadcast video programming. It is sometimes referred to by the initials DTV. Borrowing from the term "desktop publishing," this video technology and technique uses the personal computer (PC) and is affordable, portable, and versatile.

diary system   a method of determining the viewership of television programs on a local station. Members of a family record their viewing by notations in a diary.

digital communications   a process that breaks down the standard ANALOG COMMUNICATIONS signal into a series of binary numbers (0011, 0012, etc.) that are usually coded. The numbers are then transmitted digit-by-digit, decoded, and interpreted.

digital video compression   a plan for reducing the amount of information necessary to reconstruct video FRAMES at the receiving end of a transmission. The electronic signals are squeezed and thus signal capacity can be increased by factors of 8, 10, or more. The process can expand the number of channels per

satellite TRANSPONDER and create sufficient channel capacity to make DIRECT BROADCAST SATELLITE (DBS) systems practical, and it will be used in HIGH DEFINITION TELEVISION (HDTV) systems and hasten the advent of ADVANCED TELEVISION (ATV). Cable operations will be able to transmit hundreds of channels over a single FIBER OPTIC cable with subsequent economies.

digital video effects (DVE)   dazzling, sophisticated "optical" images from the surreal to the elegant, created by using video in a filmic way. Using DIGITAL COMMUNICATIONS techniques, it is possible for images to be gathered, stored, broken down, clipped, dissolved, flipped, spun, squeezed, or otherwise manipulated on a FRAME-by frame basis.

digital video recording formats new, professional VIDEOTAPE FORMATS that use the principles of DIGITAL COMMUNICATIONS as opposed to ANALOG COMMUNICATIONS recording techniques. They record video in terms of 1s and 0s. The signal is measured and expressed numerically at very frequent intervals, often at 16 million samples a second.

dimmer   a device that controls the brightness of lighting instruments in a television studio. The crew can adjust the lights and raise or lower their intensity by manipulating the levers or sliders that control the various lights.

diode   an electronic device that allows the transmission of electricity in only one direction. Diodes used to convert alternating current (AC) to direct current (DC) are called rectifiers.

direct broadcast satellite (DBS) a proposed communication service in which hundreds of signals are retransmitted by SATELLITES directly to the general public at low cost. DBS service would provide as many as 100 channels of programming to any home equipped with a small receiving antenna. Mounted on the roof or in a window, the one-to-three-foot dish would be relatively inexpensive.

New technology, including increasingly powerful satellites, DIGITAL VIDEO COMPRESSION (which promises to squeeze 10 or more signals into a single transponder), and the further refinement of the small home antennas may make DBS a reality in the 1990s.

direct response a marketing technique that relies on advertising as the sole means of selling to the consumer. Customers respond and place orders via an 800 telephone number or by returning a coupon. The home video industry often markets its product via direct mail. The procedure is sometimes called "direct mail marketing."

directional microphones a microphone that has a single pickup pattern, from the direction in which it is pointed. Directional mikes are often used on podiums for picking up one or two speakers.

Directors Guild of America (DGA) a trade union representing directors, first and second assistant directors, unit managers, and technical coordinators in theatrical motion pictures and television and in commercial production on both film and videotape.

Directors Guild of Canada (DGC) a Canadian trade union representing a number of positions in film and television production in addition to television directors. It is similar to the DIRECTORS GUILD OF AMERICA (DGA).

disconnects subscribers that have been dropped from a cable system. They have voluntarily chosen to discontinue the service or have been terminated by the operator because of a failure to pay the subscription fees.

Discovery Channel, The a BASIC CABLE SERVICE that schedules DOCUMENTARY programs in science, history, technology, nature, and travel, primarily for an adult audience. In 1991 the network acquired THE LEARNING CHANNEL (TLC) and expanded its scope even further, creating a new parent organization known as Discovery Communications.

dissolve a method used in the transition between scenes in a television production. A gradual blending of one shot into another, the technique is called a CROSSFADE in audio and lighting and a "lap dissolve" in film. A dissolve is used to indicate a slight break in continuity, and in dramatic programs it often implies a transformation in time or place.

distance education an instructional situation in which a teacher and students engage in the educational process while physically separated from one another. Older terms are sometimes used in referring to the concept, including "external education" and "independent study" as well as "home study" and "distance learning." Since the late 1960s, the word "open" has often been used with the phrase to define a situation in which an academic institution that is sponsoring distance education courses has waived its academic requirements for admission.

There is no single method of instruction in distance education. The most common techniques involve a teacher in a TV studio who presents a lesson that is transmitted live to students in an area or school district via broadcast or cable, or throughout the nation by satellite. Students can ask or respond to questions and use electronic mail, fax machines, or computers to more fully communicate with the teacher.

distant signals signals that are imported by a cable system from outside the local viewing area. The transmissions are usually from television stations located in other areas and the nonlocal programs are used to flesh out a cable system's program schedule.

distribution amplifier (DA) a piece of electronic equipment that receives a single signal input, amplifies it, and distributes it among multiple outlets. A version of the device is used in a cable system to boost the signals just prior to their reception at subscribers' homes. The DA is sometimes called a "feeder amplifier" or "live extender amplifier."

distributor See SYNDICATOR and WHOLESALER.

Division of Educational Media Management (DEMM) a division of the ASSOCIATION FOR EDUCATIONAL COMMUNICATIONS AND TECHNOLOGY (AECT). It is composed of professionals in the field of AUDIOVISUAL COMMUNICATIONS management.

Division of School Media Specialists (DSMS) a division of the ASSOCIATION FOR EDUCATIONAL COMMUNICATIONS AND TECHNOLOGY (AECT) that seeks to improve and promote communication among K–12 AUDIOVISUAL COMMUNICATIONS school personnel.

Division of Telecommunications (DOT) a division of the ASSOCIATION FOR EDUCATIONAL COMMUNICATIONS AND TECHNOLOGY (AECT) that seeks to improve education through the use of television and video.

docudrama a program FORMAT that consists of shows that are usually fictional re-creations of real events or dramatized versions of the lives of historical personalities. The first part of the term is derived from DOCUMENTARY and the approach is similarly serious in tone and style. Although the programs take considerable liberties with history or actual events, they do offer a sense of realism.

documentary   a television and film FORMAT concerned with relating or documenting an actual event or situation. The term is an offshoot of the French word *documentaire,* for travel films. It was initially used by the Scottish film maker John Grieson, who called it "a creative treatment of actuality."

Documentaries are usually supported by a meticulous examination of the facts and have a statement to make or a theme or purpose.

dolly in/dolly out   camera movements used in both television and film production, which involve moving the camera toward or away from an object or subject. Although the effect is similar to zooming in or out, the perspective of the shot is less distorted when the camera physically moves toward or away from a subject. Some directors use the terms "push in" or "pull back" or "dolly back" to achieve the same effect. The terms, ironically, were derived from the dollies that support TRIPODS but few professionals attempt to use these rather unsteady mounts to dolly in or out on the air today. (See also FRAMING and ZOOM LENS.)

donut   a commercial that has a blank section (hole) in the middle. While the material at the beginning and end of the commercial remains the same, new and timely information about particular products or bargains is inserted in the middle.

double billing   an unethical and illegal method of charging a national manufacturer of a product or service more than the cost listed on the local RATE CARD for COMMERCIAL TIME or more than the established costs of print advertising. Sometimes the national firm is billed twice for the same SPOT.

double pumping   a program scheduling technique that aims to call attention to a new series by scheduling the premiere episode of the series twice. The term is a reference to the need to pump the handle of some water pumps at least twice before the water begins to flow.

downgrade   a circumstance occurring when subscribers to a cable system choose to discontinue certain channels from their PAY (PREMIUM) CABLE SERVICE or BASIC CABLE SERVICE, thereby descending to a smaller number of program choices. Downgrades often occur when TIERING is reintroduced and subscribers reevaluate their overall cable service. (See also UPGRADE.)

downlink   the entire sky-to-ground SATELLITE system. It includes the transmitting TRANSPONDERS on the satellites as well as TELEVISION RECEIVE ONLY (TVRO) dishes and attendant receiving electronics gear that is part of the EARTH STATION. The term is often specifically (and erroneously) applied only to the TVRO. The term is also often used (correctly) to describe the process in which a signal is "downlinked" from a satellite. (See also DIRECT BROADCAST SATELLITE [DBS] and TELEPORTS.

downtime a period when a television or film studio is inoperable and the facility and its equipment are idle. The circumstance is sometimes forced upon an operation because of equipment malfunction, but every studio schedules downtime occasionally to allow for equipment maintenance and housekeeping.

dramedies a term that loosely describes dramatic programs that have some comedic elements. The shows are usually an hour in length and are often continuing series.

drop line See CABLE DROP.

dropouts a phenomenon of videotape recording and playback that occurs when the spinning recording or playback heads pass over a dead, bare, or dirty spot on the tape and black or white lines or spots appear momentarily in the picture. The glitches occur at places where the magnetic particles have flaked off or where dirt or dust covers that portion of the tape, interfering with signal pickup.

dual deck VCR See VCR-2.

dubbing a slang term in audio and video production that refers to a duplication procedure. The ability to reuse tape and the ease of replication is one of the major advantages of using magnetic tape rather than film in television production. The number of copies that can be made is virtually unlimited. Dubbing can be accomplished from one VIDEOTAPE FORMAT to another—for example, from a larger format to a smaller one (2-inch to 1-inch) or from smaller to larger (3/4-inch to 1-inch). That process is called "bumping up" or "bumping down."

The term dubbing is also used to refer to the process of substituting one audio track for another. LIP SYNCHRONIZATION is often used in film and television production to replace one person's voice with another's. The new version is said to have a "dubbed" soundtrack.

duopoly rules FEDERAL COMMUNICATIONS COMMISSION (FCC) rules dating from 1943 and 1944 that limited any licensee to no more than one type of station in any MARKET. Under these rules no licensee could own more than one station (AM–FM or TV) in a single locality.

In 1989 the Commission relaxed the duopoly rules to cover only a principal city contour rather than the wider AREA OF DOMINANT INFLUENCE (ADI), thus allowing some stations that are jointly owned to be located closer together. In 1992 the FCC moved to relax the rules further by permitting single ownership of three to six radio stations in any market, dependent on market size.

dupe See DUBBING.

DVI initials standing for digital video interactive, a new format developed from the evolution of the COMPACT DISC (CD), CD-ROM, and CD-I technology. The DVI device combines audio and video material in one DIGITAL COMMUNICATIONS format. It is sometimes referred to as IVD, for interactive video disc.

The technology uses the process of DIGITAL VIDEO COMPRESSION in the encoding of a five-inch (in diameter) disc and the restoration of the information in the playback mode. It is displayed on a computer screen.

DVS See DESCRIPTIVE VIDEO SERVICE.

dynamic microphone a pressure-sensitive device with a diaphragm connected to a wire coil that moves dynamically in a magnetic field in response to sound waves. Such a mike can tolerate very high sound-pressure levels without creating distortion. The mike can have an OMNIDIRECTIONAL, CARDIOID, or UNIDIRECTIONAL pickup pattern and is often a LAVALIERE or SHOTGUN MICROPHONE.

E

Early Bird satellite one of the pioneer communication SATELLITES that was the first to make live television across the Atlantic ocean possible. Officially known as Intelsat I, the "bird" was launched in April 1965 by the National Aeronautics and Space Administration (NASA) for the AMERICAN TELEPHONE AND TELEGRAPH COMPANY (AT&T). The first satellite in GEOSYNCHRONOUS ORBIT, it was positioned over the Atlantic Ocean and carried one television channel and 240 telephone circuits.

early fringe a DAYPART that is most often considered the period between 4:00 P.M. and 6:00 P.M. eastern standard time (EST) Monday through Friday. Some SYNDICATORS, stations, and ADVERTISING AGENCIES maintain that early fringe begins at 4:30 P.M. and continues into PRIME-TIME ACCESS at 7:30 P.M.

earth station an overall generic term describing the terrestrial TRANSMITTER, ANTENNA, and associated equipment used to transmit or receive signals from a communications SATELLITE. The definition is often broadened to include all of the ground electronic equipment associated with satellite transmission and reception, including the buildings that house the gear.

Eastern Educational Network (EEN) a private, nonprofit regional PUBLIC TELEVISION (PTV) network providing a range of public and INSTRUCTIONAL TELEVISION (ITV) programming services to its member stations. Its membership consists of 31 stations largely in the northeastern part of the United States. EEN also owns and manages the INTERREGIONAL PROGRAM SERVICE, a national program SYNDICATION service. (See also GROUP BUY.)

Echo I satellite the first successful communications SATELLITE, launched by the National Aeronautics and Space Administration (NASA) on August 12, 1960. It carried no electronic receiving or transmitting gear but was simply a 100-foot balloon, which was placed in a random, elliptical orbit. NASA used the device to experiment with sending signals, which were bounced off its surface and received back on earth.

EDISON an electronic information and communications system for public telecommunications personnel, operated by the CENTRAL EDUCATIONAL NETWORK (CEN). Using personal computers (PC), participants from PUBLIC TELEVISION (PTV) stations, agencies, and organizations can send private electronic mail, access members' data bases, view a national calendar of events in the field, or participate in a discussion. (See also LEARNING LINK.)

EDTV systems See ENHANCED NTSC SYSTEMS.

educational communications See AUDIOVISUAL COMMUNICATIONS.

Educational Film Library Association (EFLA) See AMERICAN FILM AND VIDEO ASSOCIATION (AFVA).

educational media See AUDIOVISUAL COMMUNICATIONS.

educational technology See AUDIOVISUAL COMMUNICATIONS.

educational television (ETV) a broad term that describes the use of television to inform, enlighten, and instruct. Most people use the phrase in connection with programming of a noncommercial nature that illuminates ideas and brings new concepts and experiences to the viewing audience.

The FEDERAL COMMUNICATIONS COMMISSION (FCC), in its SIXTH REPORT AND ORDER, allocated some channels in 1953 for the specific use of noncommercial broadcasting. The stations that occupied these channels were "ETV stations." That enterprise is still called "noncommercial educational broadcasting" by the FCC, even though the PUBLIC BROADCASTING ACT of 1967 labeled the industries "public radio" and "public television (PTV)."

Today, the term "PTV" has replaced "ETV" in popular usage in the United States and the older term is considered obsolete. The term "educational television," however, is often loosely used as an umbrella term to encompass all systematic as well as nonsystematic educational programming in many other countries.

Educational Television Facilities Act an amendment to the COMMUNICATIONS ACT OF 1934 that played a critical role in the development and growth of EDUCATIONAL (later public) TELEVISION (ETV) in the United States. The result of an intensive campaign by the National Association of Educational Broadcasters (NAEB), the Act was signed into law by President Kennedy on May 1, 1962. Using funds from the Act, the number of ETV stations doubled in five years.

effective competition rules FEDERAL COMMUNICATIONS COMMISSION (FCC) rules concerning cable systems that come under the regulatory control of local FRANCHISE authorities. The CABLE COMMUNICATIONS POLICY ACT OF 1984 permitted cable rate regulation only in local communities where there was no "effective competition." The FCC determined that a cable system had "effective competition whenever at least three unduplicated stations serve the cable community." From a practical standpoint, almost no communities were not covered by at least three stations, and as a result, the interpretation effectively deregulated cable rates to virtually all subscribers.

In 1991 the Commission changed the rules. A system would only be exempt from local rate regulations if the franchise area was served by six or more nonduplicative broadcast signals. In another standard, a system would be exempt if its market was served by another cable system or another multichannel video provider that was available in 50 percent of the homes and had at least a 10 percent penetration.

effective radiated power (ERP) the power (expressed in kilowatts) of a station's visual signal. In the United States, television stations are author-ized by the FEDERAL COMMUNICATIONS COMMISSION (FCC) to operate at a certain power. In order to avoid interfering with other electronic communications, stations are limited in the amount of power that can be emitted from their TRANSMITTERS and ANTENNAS. A station may have 149 kw visual power and 29.5 aural power, but its ERP is expressed as 149 kw.

EFP electronic field production. The term was coined in the 1970s by equipment manufacturers to identify their new, smaller, production gear and the accompanying techniques that allowed easier REMOTE productions. Today, the same gear is used for ENG operations, but EFP production techniques are usually more similar to studio work. There is a careful setup and production plan and EFP work requires a larger crew and involves more logistics than a simple ENG, one-camera operation.

8mm video format a VIDEOTAPE FORMAT that can be obtained as a separate deck, but is used more often as a component part of a CAMCORDER. The width of the metal tape (about one-third inch) is the same as the film that dominated the home-movie market for years. Introduced in the early 1980s, the device can be set to record for either two or four hours. Most of the machines can also play back the recorded videocassettes. While this home video unit produces superior pictures (often better than conventional VHS camcorders), the quality of the pictures is not sufficient for broadcast purposes.

Electra a broadcast TELETEXT operation consisting of a one-way information service. It contains pages of digital information from a central computer data base, emanating from television station WKRC-TV in Cincinnati. Electra information is also fed to and transmitted by SUPERSTATION WTBS to cable systems, and some television stations pick up and retransmit the information from the WTBS signal. The Electra operation is based on the WORLD STANDARD TELETEXT (WST) system.

electronic data interchange (EDI) a buzzword in retailing and the home video industry relating to a process that relies on computers for electronic messaging. EDI techniques can eliminate hours of paperwork and errors in ordering, processing, and invoicing products as well as in keeping track of retail sales. Searching for titles and ordering from WHOLESALERS are simplified by EDI, and retailers can control the inventory of their own and any sister store. Special orders can be made while the customer is in the store and orders can be made after business hours via computer.

Electronic Industries Association (EIA) a trade association and its Consumer Electronics Group (CEG) that includes the majority of manufacturers of CONSUMER ELECTRONICS products. The EIA/CEG tracks the production, sales, and inventories of the products, prepares reports and industry-wide research studies, and represents its members before Congress and the public.

Electronic Media Rating Council (EMRC) an organization that establishes and monitors the standards for RATINGS surveys. Its membership consists of industry TRADE ASSOCIATIONS, broadcast and cable networks, and owners of electronic media companies.

electronic photography See KODAK STILL-PICTURE PROCESS, STILL VIDEO CAMERAS, and STILL VIDEO PRINTERS.

electronic shopping See HOME SHOPPING NETWORKS I AND II (HSN), QUBE, and QVC NETWORK INC.

Electronic Technicians Association, International (ETA-I) a nonprofit association whose members are electronic technicians and educators in electronics and satellite technology. The ETA-I maintains a book and videotape library, which is used to practice for certification exams, and administers such exams.

Electronics Representatives Association (ERA) a nonprofit trade association whose membership consists of independent sales representatives. The reps sell electronic components and materials, including AUDIOVISUAL COMMUNICATIONS and video products for institutional and home use.

emcee often written as MC (the initials for "master of ceremonies"). This unique phrase is VARIETESE for the host on a one-time-only (OTO) program or variety or GAME SHOW. The role originated in vaudeville days and in minstrel shows, where a "Mister Interlocutor" was the centerpiece of the action. (See also SPECIALS.)

emergency broadcast system (EBS) a nationwide system by which stations can tie into a specific FREQUENCY that will broadcast information during a national emergency or crisis. Although it can be used by state and local officials, it can only be activated nationally in the 582 EBS operational areas by the White House. It is designed to enable the president to speak to the nation within 10 minutes of submitting such a request. The system is tested periodically on radio, television, and cable networks and stations.

Emmy awards awards given annually, in the form of gold statuettes, to recognize television broadcast accomplishments. They are voted on by the members of the NATIONAL ACADEMY OF TELEVISION ARTS AND SCIENCES (NATAS) and the ACADEMY OF TELEVISION ARTS AND SCIENCES (ATAS).

The awards for PRIME-TIME programs (conferred by ATAS) are given in various (often changing) categories such as Best Actor in a Supporting Role in a Special or Best Music in a Miniseries. The award ceremonies for the prime-time entertainment programs are glamorous events, telecast nationally from Hollywood. The awards given for daytime programs and in sports and news are administered by NATAS. The name Emmy is a derivation of the industry word "immy," a slang term for the IMAGE ORTHICON TUBE.

encryption See SCRAMBLING.

end rate the final rate paid by an advertiser or ADVERTISING AGENCY for COMMERCIAL TIME on a cable or television operation. Because so many combinations and discounts are available on RATE CARDS for different SPOTS, the rates vary a great deal. After all negotiations are completed and discounts are applied, the end rate is the actual rate paid by the advertiser. (See also PREEMPTABLE RATES, RUN-OF-SCHEDULE [ROS] RATES, SPOT [CONTRACT] and STANDARD RATE AND DATA SERVICE.)

ENG electronic news gathering, a technique used to cover news events in the field. The term is also applied to the technical gear and tape equipment used for such purposes. The method replaced the traditional 16mm film news coverage in the early 1970s. ENG gear and personnel are an integral part of the SATELLITE NEWS GATHERING (SNG) process. (See also EFP, PORTAPACK, and TIME BASE CORRECTOR.)

enhanced NTSC systems types of proposed ADVANCED TELEVISION (ATV) systems that use an approach that modifies the existing television technical standards in a modest manner. The enhanced systems seek to improve (or "enhance") the current NTSC system with better picture quality, but the method does not increase the number of scanning lines.

The systems will, however, improve the ASPECT RATIO and will work within the current 6-MEGAHERTZ (MHZ)-channel. They are described as "NTSC-compatible" in that the basic signal can be received on current TV sets, but in order for the picture improvements to be seen, a new television set would be required. This approach is sometimes called "enhanced definition television" (EDTV). Three variations of such a system have been proposed. They are viewed as an evolutionary step toward HIGH DEFINITION TELEVISION (HDTV) using digital technology.

equal employment opportunity (EEO) FCC rules FEDERAL COMMUNICATIONS COMMISSION (FCC) rules that encourage the employment of minorities. Federal agencies include in their classifications of minorities, people who are black, Asian-American, Pacific Islanders, American Indian, and Hispanic. Women (of all races) are also included under EEO Rules.

Any applicant for a CONSTRUCTION PERMIT (CP) for a broadcast operation that will employ more than five people fulltime must file an EEO plan with the FCC. After the station obtains an FCC LICENSE, it is required to file an annual EEO report with the Commission. Cable systems with more than six employees must also file annual EEO reports with the Commission. The report must detail what the operation has done in the EEO area. The FCC regulations also contain a series of detailed requirements related to the number of women and minority employees that must be on staff in relationship to the number of them in the local labor force.

Equal Time (Opportunity) Rules Section 315 of the COMMUNICATIONS ACT OF 1934 as amended. The rules state that any broadcast station that permits a legally qualified candidate for public office to use its facilities must afford an "equal opportunity" for use of the same facilities to all other candidates for that same office.

Many professionals use the term "equal time" when referring to the rights of political candidates on broadcasting stations. The correct phrase, however, is "equal opportunity" and the difference in the two phrases is important. A candidate appearing for an hour at 8:00 P.M. in PRIME TIME has a distinct advantage over a candidate who appears at 8:00 A.M. That candidate may be given the "equal time" of one hour, but not an "equal opportunity," because the early-morning broadcast will not reach as many people.

There are some exceptions to the equal opportunity rules, as interpreted by the FEDERAL COMMUNICATIONS COMMISSION (FCC). Appearances by candidates on any

1. *bona fide* newscast,
2. *bona fide* news interview,
3. *bona fide* news documentary (if the appearance of the candidate is incidental to the presentation of the subject or subjects covered by the news documentary), or
4. on-the-spot coverage of *bona fide* news events (including but not

limited to political conventions and activities incidental thereto) are exceptions to the rules and are not covered by the requirements of Section 315. FCC interpretations of what constitutes a *bona fide* exception have been left largely to the stations. Equal opportunity rules also apply to cable systems, but only to those channels that originate programming. (See also FAIRNESS DOCTRINE and POLITICAL EDITORIAL RULES.)

ESPN Inc. a cable network, founded as the Entertainment and Sports Programming Network in 1978, that has become the largest in the United States in terms of subscribers. In 1985, the original name was dropped and a new logo and the name ESPN Inc. were adopted.

European Broadcasting Union (EBU) a nonprofit organization that encourages cooperation in radio and television via treaties, workshops, and seminars. The organization also operates Eurovision, an interconnection service that links European and Mediterranean broadcasting networks for the coverage of special events such as coronations and soccer championships. The association is not a trade union, despite the implication of its name.

Eurovision See EUROPEAN BROADCASTING UNION (EBU).

evergreen a prerecorded videocassette that is a strong, constant seller year after year at the video retail store. Movies such as "Gone With the Wind" and "Casablanca" are evergreen titles and are a major part of the CATALOG PRODUCT of their respective PROGRAM SUPPLIERS.

EVR an obsolete electronic video recording device that was a miniaturized film cassette system played through a television set. Introduced in 1970 by CBS, the unit played a seven-inch diameter cartridge of 8.75mm film containing one hour of programming, through a TV set. The cartridge was produced by exposing film to a beam of electrons using an electron beam recorder. Unfortunately, such recorders produced masters that varied in quality. Because there was no mass market yet, only short duplicating runs were made, and the result was that each 20-minute duplicate cartridge cost between $50 and $100 to produce.

At about the same time, the SONY CORPORATION introduced its 3/4-INCH U(EIAJ) VIDEO FORMAT, which could record at home on blank videocassettes, cost only $30 for one hour of programming and could be easily duplicated. It could also play back. CBS saw the handwriting on the wall and abandoned EVR in August 1971, taking a loss of a reported $33 million.

exclusivity the singular right to use a particular product that excludes others from its use. In television, cable, and video, exclusivity is most often sought and granted in the area of programming. SYNDICATORS who pick up programs produced by an INDEPENDENT PRODUCTION COMPANY will seek exclusive distribution rights before they begin to market the product. Stations look for exclusive rights for any syndicated program they purchase, seeking to prevent any other station in their area from carrying the same series. Broadcast networks insist on exclusive rights for sporting events to restrict cable networks and local operations from competing for the sports viewer with the same show. Cable networks seek exclusive cable rights for motion pictures to air on PAY (PREMIUM) CABLE networks, and home video WHOLESALERS look for some territorial exclusivity in which they can sell prerecorded videocassettes from a

PROGRAM SUPPLIER to home video retail stores. In an attempt to alleviate the problem of program duplication in many markets, Congress reinstituted SYNDICATION EXCLUSIVITY RULES in 1988.

ExtraVision one of the U.S. survivors of the battle for the establishment of a one-way, broadcast, TELETEXT information system. Based on NORTH AMERICAN BROADCAST TELETEXT STANDARD (NABTS), ExtraVision is programmed by a company in Washington, D.C. for CBS and became available in 1983 for stations affiliated with that network.

F

f-stop a measure of the amount of light entering the lens of a still, film, or television camera. The f stands for fixed. Numbers are etched on the IRIS ring on the front of the lens, denoting the extent to which the iris is closed or open. The lower the number, the larger the iris opening with more light allowed into the camera, and the "faster" the lens. F-stop numbers are the product of a mathematical formula where f is equal to the FOCAL LENGTH of the lens divided by the diameter of the lens.

facilities the physical aspects of a television station or production com-

pany. The term is applied more specifically to technical and production gear, including DISTRIBUTION AMPLIFIERS (DA), CAMCORDERS, CHARACTER GENERATORS, VIDEOTAPE FORMATS, and all other production and engineering equipment, and is often expanded to include the station's TRANSMITTER and EARTH STATION installations. The equivalent of facilities in the cable television industry is PLANT.

fade-in/fade-out a production technique that creates a major change in the content of a television or film program. Fades are accomplished by using the levers (faders) on a

SWITCHER to gradually increase the video signal from black to a visible picture (fade-in) or by reversing the process by fading gradually to black (fade-out). A fade-in begins a program, scene, or act, and a fade-out has the effect of a descending curtain in the theater, signaling the end or conclusion. Fade-ins and fade-outs are also frequently used in the audio portion of a program, when the volume of the sound of dialogue or music gradually increases from inaudible to full volume or vice versa. (See also CROSS FADE, CUT, and DISSOLVE.)

fair use doctrine aspects of the COPYRIGHT ACT OF 1976 that seek to balance the rights of the creator of an intellectual property with the needs of society. Limited use of copyrighted materials without a fee or permission for noncommercial purposes is deemed to fall within fair-use guidelines. The Act lists four keys to be considered in determining whether the fair-use doctrine is applicable:

1. the purpose and character of the use, including whether such use is for commercial or nonprofit, educational purposes.
2. the nature of the copyrighted work
3. the amount and (substantially) the portion used in relation to the copyrighted work as a whole
4. the effect of the use upon the potential market for or value of the copyrighted work.

The law gives six examples of fair use: (1) criticism, (2) comment, (3) news reporting, (4) teaching, (5) scholarship, and (6) research. (See also OFF-AIR VIDEO RECORDING and PUBLIC PERFORMANCE OF COPYRIGHTED VIDEO MATERIAL.)

Fairness and Accuracy in Reporting (FAIR) a nonprofit organization that is known largely for its study of television news bias. Its criticisms of television programming are often from a liberal perspective and thus contrast with the analysis and views of ACCURACY IN MEDIA (AIM).

fairness doctrine FEDERAL COMMUNICATIONS COMMISSION (FCC) regulations and sections of the COMMUNICATIONS ACT OF 1934 that formed the philosophical basis for all broadcasting on controversial issues of public importance in the United States for many years. The principle of the fairness doctrine was based on the idea that a broadcaster had to give some airtime to the discussion of issues of public importance and that the opportunity be given for the presentation of different views.

In enforcing the doctrine, broadcasters were permitted by the FCC to determine the issues of public importance and were given discretion regarding the formats to be used and the persons who would present the opposing views. The key criterion for adherence to the fairness doctrine was providing a balanced perspective.

The FCC eventually found that the fairness doctrine was unnecessary in a world of many media channels and using a court case, abolished it in 1989 except for political-editorializing and personal-attack rules.

family viewing time a broadcast program policy adopted by the television industry in 1975. It set aside the first two hours of PRIME TIME (7:00–9:00 P.M. EST) for programs appropriate for family viewing. Hollywood directors and writers were outraged and filed suit in 1976 against the networks, alleging a violation of the First Amendment. In 1984, all parties agreed to settle the matter out of court. By that time, family viewing time had died a quiet and largely unmourned death. (See also CHILDREN'S TELEVISION ACT OF 1990.)

FBC See FOX INC.

FCC license a document issued by the FEDERAL COMMUNICATIONS COMMISSION (FCC) authorizing a company or an individual to operate a full-power radio or television station, a MULTICHANNEL MULTIPOINT DISTRIBUTION SERVICE (MMDS) system, or a LOW POWER TELEVISION (LPTV) STATION in the United States. FCC licenses are also required for other broadcast services that use the electromagnetic spectrum in U.S. broadcasting. The purpose of requiring licenses is to regulate the airwaves to avoid electronic interferences between the various services and to ensure that the broadcast stations operate in the "public interest, convenience, and necessity." (See also TRAFFICKING RULES.)

FCC lottery a FEDERAL COMMUNICATIONS COMMISSION (FCC) procedure for the purpose of processing license applications for LOW POWER TELEVISION (LPTV) STATIONS and MULTICHANNEL MULTIPOINT DISTRIBUTION SERVICE (MMDS) operations. In a 1982 amendment to the COMMUNICATIONS ACT OF 1934, Congress authorized a lottery system to select applications to be reviewed. The lotteries for those two services began in 1983 and have continued on a monthly basis.

FCC zones FEDERAL COMMUNICATIONS COMMISSION (FCC) zones with rules that govern the separation of broadcasting stations within each zone. There are three geographic zones. In Zone I, the minimum co-channel separation is 170 miles for VHF channels and 155 miles for UHF. In Zone II, the minimum co-channel separation is 190 miles for VHF and 175 miles for UHF channels. In Zone III, the separation is 220 miles for VHF and 205 miles for UHF.

Federal Communications Commission (FCC) an independent government agency, responsible directly to Congress, that is charged with regulating interstate and international communications by radio, television, telephone, satellite, and cable. Established in accordance with the COMMUNICATIONS ACT OF 1934, the FCC is composed of five commissioners appointed by the president and confirmed by the Senate for five-year terms. The commissioners supervise all FCC activities, delegating responsibilities to staff units and bureaus.

The Commission allocates spectrum space for AM and FM radio and VHF and UHF television broadcast services, assigns FREQUENCIES and CALL LETTERS to stations, and designates operating power and SIGN-ON/SIGN-OFF times. The Commission also issues CONSTRUCTION PERMITS (CP) and inspects technical equipment. When a station is built, an FCC LICENSE to operate it is issued by the Commission.

In 1966 the FCC established some regulations for cable systems, including MUST-CARRY RULES. Congress passed the CABLE COMMUNICATIONS POLICY ACT OF 1984 in October of that year, which gave the FCC more regulatory responsibilities over the cable industry. The FCC also has jurisdiction over the INSTRUCTIONAL TELEVISION FIXED SERVICE (ITFS) and the MULTIPOINT MULTICHANNEL DISTRIBUTION SERVICE (MMDS), as well as DIRECT BROADCAST SATELLITE (DBS) systems and LOW POWER TELEVISION (LPTV) STATIONS.

Federal Trade Commission (FTC) a federal agency holding responsibility for the government's regulatory role in interstate commerce and the monitoring of possible price-fixing and unfair and illegal acts and practices. The latter includes fraudulent or deceptive advertising on radio, television, cable, and home video.

feedback an undesirable effect such as the screech or howl that occurs when a microphone is placed too near a speaker. Sometimes called a "back squeal," it often occurs when a mike is placed near loudspeakers in an auditorium. Video feedback is created when the camera is aimed at a monitor that is showing the picture being shot by the camera. The resulting image is a variety of random abstract patterns that are somewhat psychedelic. Feedback also refers to an audience's response to a program or a commercial in the form of telephone calls, letters, or by means of structured FOCUS GROUPS.

feeder cables an element of a TREE NETWORK cable operation consisting of the COAXIAL CABLES that connect the cable TRUNK LINES to the CABLE DROP lines. This intermediate portion of the distribution system carries the electronic signals from large trunkcable lines to a specific area or neighborhood of homes. Sometimes called "feeder lines," they are installed underground or strung between telephone poles.

fiber optics a technology that uses light beams to carry signals from one point to another through hair-thin strands of glass bundled together in a cable. LASERS are used to generate the modulated light that travels rapidly through a glass fiber by reflecting off the inside walls of the fiber. The small cables carrying the beams of light can be buried underground, laid under the ocean, or strung on poles. Fiber optics technology has many advantages over COAXIAL CABLE, including its greater capacity which is practically infinite. In Europe the relatively new technology is sometimes known as "glass optical wave guides."

field a partial image formed by an electron gun in a CATHODE RAY television receiving tube. The gun sweeps the tube with an electronic beam from side to side and from top to bottom. This scanning process takes one-sixtieth of a second (in the NTSC standard), but scans only every other line, thus creating only one-half of the picture. The second scanning process of the alternate lines also takes one-sixtieth of a second. Together, the two interlaced fields (of 525 scanning lines) create a complete picture called a FRAME.

field strength the intensity of an electronic or magnetic field at a given point. In broadcasting, the strength of the transmission is measured in microvolts per meter at various points some distance away from the transmitting ANTENNA, to determine the extent of the station's COVERAGE AREA. In cable television, a signal-level meter measures the energy level of the cable signal at various points in the COAXIAL CABLE of the system.

Fifth Estate radio broadcasting and, by extension, television. In ancient times, there were four traditional "estates" in society. The first was the Clergy, the second the Nobility, the third the Commons, and the Fourth, the Public Press. To distinguish the new, powerful radio medium from the printed page, the term Fifth Estate was coined in the 1930s.

fill lighting a television and film lighting technique that directs broad light onto a performer, object, or area from large lights positioned at the front and sides of the scene.

Sometimes referred to as base lighting, this SOFT LIGHT is designed to complete the illumination of a scene by eliminating shadows and dark spots. (See also BACKGROUND LIGHTING, BACKLIGHTING, KEY LIGHTING, and SCOOP LIGHT.)

fill rate the portion or percentage of an order that is fulfilled for prerecorded videocassettes or videodiscs by manufacturing or duplicating firms. When an order is processed, not all of the units may be immediately duplicated and shipped, and the success in meeting that goal is usually expressed in percentages (for example, "X company has a good fill rate of 90 percent"). (See also PREBOOK.)

filler programming additional programming used to flesh out a time period. In the early days of television, programs often did not conform to the traditional half-hour or hour segments and short films filled out the time slot. As television began to produce most of its own programming, the need for filler became less important.

Filler material is used on PAY (PREMIUM) CABLE networks, where it has been given the fancy name of "interstitional" (or "continuity") programming. The short bits are used to fill out time left when a motion picture does not conform to half-hour or hour scheduling.

film chain a unit that converts motion picture film or still slides into a video signal. It allows the film to be projected into the lens of a video camera and thus be recorded on videotape or transmitted to the audience. Several projectors are positioned to project into a set of mirrors in the center of the unit, which can be changed to direct the image from any of the projectors into the lens of a video camera. The device is sometimes called a "telecine."

film clip a bit of film footage, known familiarly as a "clip," that is often used as an insert in a television production. The brief film can be introduced into a live studio program to take the viewer outside the studio. In the early days of television, all such footage was shot on 16mm film or sections were physically clipped out of longer films. Scenes were also cut from theatrical films. Today, such segments are usually shot on videotape rather than film, and the clips from motion pictures are also transferred to tape, before they are used. Although the videos are not cut from the original, the term "clip" is still used, but without the prefix "film."

film loop a short section of tape or film that is run repeatedly to produce a repetitive scene. A short (eight-foot or so) length of 16mm film is spliced into a loop and run continuously, in a never-ending circle, through the projector of a FILM CHAIN. Videotape machines can also be adapted to accommodate a tape loop.

film package an assemblage of several motion pictures marketed in syndication as a single unit under an umbrella label. These collections,

created by SYNDICATORS, sometimes contain films of a particular GENRE such as westerns, but often the individual motion picture titles are disparate and are grouped together under such broad titles as "Silver Screen Classics."

fin-syn rules FEDERAL COMMUNICATIONS COMMISSION (FCC) financial interest-syndication rules (FISR), adopted in 1970, that have limited the power of the commercial broadcast networks over program producers and distributors. The financial (fin) aspect of the regulations prevented the networks from acquiring an interest in the programs from outside producers that they put on the air. The SYNDICATION (syn) aspect also prohibited networks from the distribution of any programs they owned except to overseas markets. The rules were intended to foster diversity and encourage competition in programming.

In 1991 the Commission dropped all restrictions on the networks' ability to have a financial interest in any syndicated programming broadcast during nonprimetime hours. The rules were also relaxed somewhat for their financial participation in programs that are produced by others for PRIME TIME, and the networks are now permitted to syndicate programs produced by others, overseas. The networks are also allowed to produce up to 40 percent of their prime-time programs in-house and to syndicate those shows domestically. They can also produce programs for FIRST-RUN syndication, but such distribution must be through third parties. The new rules are being challenged by interested parties in court.

final print the end result of the film production process. All elements, including sound and optical effects such as titles, SUPERIMPOSITIONS, and DISSOLVES have been added to the print, and the film, program or commercial is completed. This print has been corrected for color, checked for quality, and is ready for transmission. (See also ANSWER PRINT and WORK-PRINT.)

first-run syndicated programs shows that are created specifically for SYNDICATION and sold or licensed to local television stations, GROUP BROADCASTERS, and to cable and other television systems in the United States and overseas. The programs are developed largely by INDEPENDENT PRODUCTION COMPANIES for an initial (first-run) broadcast on non-network operations in the United States. This type of syndicated programming is distinct from OFF-NETWORK PROGRAMS, which are series, SITCOMS, or SPECIALS originally transmitted by the major networks.

first-sale doctrine a concept that allows the first purchaser of a copyrighted work (such as a book or videocassette) to do anything with the work. The buyer does not own the copyright, but the physical work may be sold, rented, or given away. The COPYRIGHT ACT OF 1976 specifically authorizes an owner of a copyrighted copy of a work "without the author-

ity of the copyright owner, to sell or otherwise dispose of the possession of that copy."

fixed position a specific time for the broadcast of a COMMERCIAL on a television station or cable operation. The rate for a fixed position is higher than that charged for a commercial purchased at RUN-OF-SCHEDULE RATES or at a PREEMPTABLE RATE, but advertisers are willing to pay a premium rate for the desired exposure.

flags lighting accessories used to shadow or block unwanted light from certain areas. They are mounted on a stand or on the grid above the studio. Sometimes called "gobos" or "cutters," they are rectangular metal frames with black fabric stretched over them.

flashback a dramatic technique in which the logical progression of events that has been occurring sequentially in a story is interrupted by a character recalling a happening from a previous time.

flat-panel television a new television technology consisting of lightweight display screens that are barely thicker than a heavy pane of glass. They are destined to replace the bulky CATHODE RAY TUBE (CRT) and MONITOR used today. The new screens, which will be similar to those used in portable computers and tiny, palm-size television sets, use liquid crystals. They can be converted to viewing patterns by an electronic signal. Active matrix displays are used, in which the liquid crystal is paired with a large computer chip and a million or more transistors. Each transistor controls specific picture elements that, in the aggregate, produce the image. Using this technique, thin large-screen television sets that can hang on the wall should be a reality by the end of the 1990s.

flight a schedule of related COMMERCIALS on a television station or cable operation. In a "winter flight," the advertisements are broadcast over a long time period. A series of several SPOTS may be called a "12-spot flight." The term is used in advertising agencies as a shorthand way of describing multiple plays of commercials.

floor plan a diagram of the floor of a television studio as seen from above. It shows the position of set pieces, objects, talent, cameras, and other production gear.

flowthrough audience See AUDIENCE FLOW.

fluid head an inexpensive device that is used to support some PROSUMER television cameras, as well as many 16mm film cameras. It consists of a circular metal container beneath a flat piece of metal that supports the camera. Its inside components, which allow the head to PAN or TILT, are encased in oil, thus allowing smoother movements than those permitted by the more limited FRICTION HEAD mounts. (See also CRADLE HEAD.)

flyaway a very small SATELLITE NEWSGATHERING (SNG) EARTH STATION. The extremely portable unit can enable a live satellite feed from previously inaccessible places within hours.

focal length the distance between the optical center of the lens and the face of the picture tube when the lens is focused on infinity. It is measured in millimeters and determines how wide an angle can be seen by the lens. The smaller the focal length, the more area that can be viewed at any given distance. The larger the focal length, the smaller the field that can be viewed by the lens at that distance.

focus group a group of several consumers brought together to examine and discuss a particular concept or product in a controlled forum. This procedure is used by advertisers and ADVERTISING AGENCIES for qualitative, rather than quantitative, research. The groups are assembled at any stage of the development and execution of an advertising CAMPAIGN. (See also FEEDBACK.)

focus out/focus in a transitional camera or time technique that is used to dramatize a shift in reality or a flashback. A character may faint and the camera defocuses. There is a DISSOLVE to another shot, which gradually comes into focus on a dreamlike scene. When that scene concludes, the defocus/focus technique is used again to make the transition back to the original scene.

follow-up programming   a type of programming produced and aired following an initial show that addresses the same topic or issue. The PUBLIC BROADCASTING SERVICE (PBS), for example, occasionally uses the technique by scheduling a panel discussion following a controversial DOCUMENTARY. (See also WRAP-AROUND PROGRAMMING.)

footcandle (fc)   a measure of the luminance or brightness of a light. A certain level of light is required before an image can be recorded by a camera, and different cameras and lenses have different light requirements. LIGHT METERS are used to measure the brightness of a scene. (See also F-STOP and LIGHTING RATIO.)

footprint   the usable coverage area of a communication satellite. From a GEOSYNCHRONOUS ORBIT at a height of 22,300 miles, the TRANSPONDERS on a satellite can be made to "see" a certain area of the earth's surface, according to the width of the beam transmitted.

Ford Foundation   a philanthropic foundation that is cited by many as the father of EDUCATIONAL (now public) TELEVISION (ETV) in the United States. The foundation helped obtain the channel allocations for noncommercial television from the FEDERAL COMMUNICATIONS COMMISSION (FCC) in 1951–52 and contributed funds to create National Educational Television (NET), the PUBLIC BROADCASTING SERVICE (PBS), and individual stations. It ceased its activities in noncommercial television in 1977.

format (programming)   the form, makeup, content, style, and organization of a program, but not its subject matter.

In television, the format can be of a game, comedy, talk, panel, interview, dramatic, variety, or musical nature. It consists of the shape, size, and placement of physical and thematic elements and their arrangement within the parameters of the framework of a program. Format differs from program GENRE, which refers to the basic content of a program, such as DOCUDRAMA, SOAP OPERA, or SITCOM. Some genres such as the last two, however, have become so familiar and common that they have become formats themselves.

The term format is used differently in radio, where it describes the type of program schedule on a particular station, such as MOR Music, Call-in Talk, or Top 40. (See also REFORMATTING and SPLIT-FORMAT PROGRAMMING.)

format (recording)   See VIDEOTAPE FORMATS.

Fox Inc.   a unit of the NEWS CORPORATION LTD, Fox Inc. is a major television corporation that has, in turn, other divisions. Fox Television Stations Inc. owns stations in Los Angeles, Washington, D.C., Chicago, New York, Dallas, Houston, and Salt Lake City. The company also owns and operates the Hollywood motion picture megastudio,

20th Century-Fox; a home video arm, Fox Video; a video LABEL in a partnership called CBS-Fox Video; and two television production and SYNDICATION units. The company also operates the Fox Broadcasting Company, which has introduced new initials into the television NETWORK universe: FBC. The fourth network was launched in 1986 by attracting previously INDEPENDENT STATIONS. The FBC stations reach more than 90 percent of the country.

fragmentation   the increasing number of listening or viewing subdivisions in the mass audience. It is the result of the increasing number of independent stations, cable channels, MULTICHANNEL MULTIPOINT DISTRIBUTION SERVICE (MMDS) and LOW POWER TELEVISION (LPTV) STATIONS, home video titles, and the many new technological devices. (See also NARROWCASTING.)

frame   a complete television picture on a CATHODE RAY TUBE (CRT), consisting of 525 scanning lines (in the NISC standard). It is composed of two combined FIELDS. A frame is created by an electron gun scanning first the odd-numbered field of 262-½ lines and then the even-numbered field of 262-½ lines, each in one-sixtieth of a second.

framing   the process of composing a camera shot in television and film. The perspective of the television camera is always in the rectangular 3:4 ASPECT RATIO. The composition of the image, however, is determined by the camera operator. The compo-sition is accomplished by changing the CAMERA ANGLE, PANning, TILTing, or DOLLYing IN or OUT.

franchise (cable)   the right to construct and operate a cable system in the United States. Congress has defined a franchise as "an initial authorization or renewal thereof . . . issued by the franchising authority . . . which authorizes the construction or operation of a cable system." No cable operator can provide service without a franchise. Under the requirements of a local ordinance developed by the community, cable firms compete to obtain a franchise by submitting bid proposals that detail their financial and technical background, the number of channels they will provide, and how much they will pay the community for the right to hold the franchise.

The city council or other franchise authority may award "one or more franchises within its jurisdiction," but in practice, only one franchise is usually granted because of economic reasons. A franchise is usually granted for 10 or 15 years and is renewable. (See also CERTIFICATE OF COMPLIANCE and OVERBUILD.)

franchise (home video)   The right to market video products under a national name at the retail level in a particular geographic area. The granting of a franchise is made by a company that owns a specific name and merchandising image and sells it to a local operator. The parent company is registered in every state where it has granted franchises, and its operations are regulated by the

FEDERAL TRADE COMMISSION (FTC) in order to protect franchise holders.

freeze   the period of time (1948 to 1952) during which no new television stations were licensed. After the conclusion of World War II, the FEDERAL COMMUNICATIONS COMMISSION (FCC) was swamped with applications for new stations. In 1948, it ceased the processing of all applications for CONSTRUCTION PERMITS (CP). From then until 1952, the FCC held hearings and commissioned studies to develop a major plan that would establish engineering standards and allocate channels throughout the United States. The freeze was lifted on April 14, 1952, with the issuance by the FCC of its SIXTH REPORT AND ORDER. (See also ALLOCATIONS.)

frequency (advertising)   the average number of times a typical household or an individual viewed a given television program, station, or commercial during a specific time period. A MEDIA PLAN developed by an ADVERTISING AGENCY always contains the frequency objectives of the CAMPAIGN.

frequency (engineering)   the rate at which electromagnetic energy waves repeat themselves. Some waves vibrate rapidly and are said to have high frequencies, while others oscillate slowly and have low frequencies. The frequency of a wave is measured by the number of times a signal vibrates in a second. It is expressed in HERTZ (HZ), which by international agreement is defined as one cycle per second. Radio uses relatively low frequencies, while television broadcasting uses high frequencies and MICROWAVE RELAY technology and communications SATELLITES use an even higher part of the spectrum. The spectrum, however, is limited by the natural properties of wave propagation. In the United States, the FEDERAL COMMUNICATIONS COMMISSION (FCC) assigns frequencies for various purposes.

frequency modulation (FM)   a process that involves the alteration of the amplitude, phase, or FREQUENCY of a signal. It is a method by which signals are changed prior to transmission.

In FM, the frequency of the base (or carrier) signal is varied by the modulating signal. The stronger the modulating signal, the greater the deviation from the base frequency. Technically more complicated than AMPLITUDE MODULATION, FM is not as subject to electrical interference and is used in FM radio and for the audio signal on TV transmissions. (See also ANALOG COMMUNICATIONS, ANTENNA, and TRANSMITTER.)

frequency response   the reaction of an electronic device to signals at various FREQUENCIES. Most often used in measuring the quality of audio gear, the term describes the relationship between the gain (or loss) and the frequency in a microphone, amplifier, or speakers, and therefore the fidelity of the signal.

friction head   an inexpensive device that supports small television and film cameras. It features various locks that can be tightened or loosened to curtail or allow camera movement. Friction helps accomplish smoother PANS and TILTS, but the device does not counterbalance the camera, making such movements somewhat unstable. (See also CRADLE HEAD and FLUID HEAD.)

fringe area   the listening or viewing territory that is at the extreme limits of a transmitted broadcast signal. Viewers who are at the periphery of the reception of the broadcast are said to be in a fringe area. The reception of the signal in that location is very poor.

frontloading   the practice of placing the bulk of COMMERCIALS and/or print advertisements at the beginning of an advertising CAMPAIGN.

full-barter syndication   the practice of exchanging free programming (along with COMMERCIALS from various national advertisers) for COMMERCIAL TIME on local cable or television operations. The local operator receives free programming but is not allowed to sell any SPOTS within the program. (See also BARTER SYNDICATION, CASH/BARTER SYNDICATION, and SPLIT-BARTER SYNDICATION.)

full-service advertising agency   an ADVERTISING AGENCY that offers complete services to clients. Such an agency provides all of the traditional advertising services, including research and the creation and placement of COMMERCIALS, as well as nonadvertising functions such as the preparation of annual reports, publicity materials, and trade show exhibits. (See also BOUTIQUE AGENCY.)

G

g–Star satellites See GTE SPACENET CORPORATION.

Gabriel awards   awards bestowed annually by the UNDA–USA, the American branch of the International Catholic Association for Radio and Television, to honor commercial or religious broadcasters for programs that creatively treat issues concerning human values. The awards are given in a number of categories including entertainment, information, religion, children's programs, and PUBLIC SERVICE ANNOUNCEMENTS (PSAs), and also for outstanding achievement by a television station.

gaffer   the chief electrician on a motion picture set. It is used in television only in the production of long-form dramatic programs such as MINISERIES or MADE-FOR-TV MOVIES. The term may have originated in early carnivals in Europe where an individual was in charge of herding or "gaffing" people into the tent. (See also BEST BOY and GRIP.)

gaffers' tape a unique and ubiquitous tape used in television and film production. It is two inches wide and made of vinyl-coated cotton cloth with an adhesive backing of synthetic rubber-based resin. It is available in 12 colors and has a tensile strength of 50 pounds per inch. Because of its stubborn strength and versatility, it is used for everything in a television studio.

game shows a television show format in which various kinds of games are played by people in a studio setting. Game shows often feature a charming host who directs the proceedings. The prizes consist of merchandise given to the show by a company or advertiser in order to gain exposure and PLUGS for its products.

gel short form of "gelatin," a translucent filter material used with lighting instruments to change the color, the quality, or the amount of light on a scene in the theater or in television or film production. Sometimes called a "media," this fade-resistant, cellophane-like material comes in various colors and is mounted in a frame that is attached to the front of the lighting instrument.

Gemeaux awards television programming awards presented by the ACADEMY OF CANADIAN CINEMA AND TELEVISION. The awards honor excellence in various categories in French-Canadian programming and are the equivalent of the EMMY AWARDS in the United States.

Gemini awards programming awards presented by the ACADEMY OF CANADIAN CINEMA AND TELEVISION. The awards honor excellence in various categories in English-Canadian programming and are the equivalent of the EMMY AWARDS in the United States.

General Electric Company see NATIONAL BROADCASTING COMPANY (NBC) and RADIO CORPORATION OF AMERICA (RCA).

generation a term alluding to the number of times a videotape has been duplicated from the master tape in the DUBBING process. A first-generation dub is the copy that was duplicated from the original tape and a second-generation dupe is one that has been transferred from a first-generation copy.

Genie awards film awards presented each year by the ACADEMY OF CANADIAN CINEMA AND TELEVISION. The awards honor excellence in various categories in theatrical film and are the equivalent of the OSCAR AWARDS in the United States.

genre a content-based category of television programming. DOCUDRAMAS, SOAP OPERAS, and SITCOMS are examples of program genres, as are children's programming and RELIGIOUS PROGRAMMING. Although soap operas and sitcoms are technically types of program genres, they are often called FORMATS because their style has become so familiar and accepted in the industry. Action/adventure, travel, sports, science fiction, nature, mystery, horror, animal, and educational or instructional shows are other examples of program genres.

geosynchronous orbit an imaginary circle in space that is precisely 22,300 miles above the equator. A satellite in such an orbit revolves around the earth at the same angular velocity as that of the earth's rotation around its axis. Although the satellite is moving at nearly 7,000 miles per hour, its centrifugal force nullifies the gravitational force from the earth and thus the satellite appears to be stationary. The satellite is then said to be in geosynchronous (or geostationary) orbit.

ghosts secondary images on a television screen just to the right or left of the main image. Broadcast signals reflected off mountains or tall buildings create echoes that arrive at television sets microseconds after the primary signal. The over-the-air main signal and the secondary reflection are decoded at the same time, resulting in a second image.

gigahertz (GHz) a unit of FREQUENCY in the electromagnetic spectrum designating one billion cycles per second. (See also HERTZ [HZ].)

global village the effect that all of the new MEDIA (and particularly television and SATELLITE communica-

tions) have had on the world. Modern technology and instantaneous communications have, in effect, made national boundaries obsolete and reduced the world to a smaller community where common experiences, events, and ideas are shared by everyone. The term was coined by Marshall McCluhan.

Grade A and B coverage See COVERAGE AREA.

grandfathering a ruling that exempts an organization from new governing rules because the group was in existence prior to the regulations or was operating legally under previous rules and regulations. Most recent federal, state, and local laws concerning television and cable have grandfathering clauses.

gray scale a chart of 10 steps from pure white to velvet black. It consists of vertical bars of different shades of gray. It is used by engineering, staging, and graphics personnel to test and balance the brightness and contrast in the television image and to ensure that one object will stand out from another. (See also COLOR BARS.)

grazing the process of changing television channels rapidly in search of new or different entertainment or information.

grip a stagehand on a motion picture set. The term is usually used in television only in the production of a MADE-FOR-TV MOVIE or MINISERIES. A grip handles most of the nontechnical gear (except lighting devices) and sets up and dismantles scenery, operates a crane or dolly, and sets up stands and supports for lighting equipment.

The term is said to have originated to describe the deckhands on British ships, who often had to grip the rails to steady themselves and do their work in heavy seas. The term moved into the world of theater where it was applied to the stagehands who had to grip a piece of scenery in order to move it. (See also BEST BOY and GAFFER.)

gross rating points (GRPs) the sum of all RATING points for a series of programs or COMMERCIALS that are earned in a particular period of time. The ratings reflect the total number of households or people tuned into the programs or SPOTS. The people or households may be counted more than once during the particular time period. Thus, GRPs measure duplicated audiences, in contrast to numbers represented in a CUME. (See also COST PER RATING POINT.)

group broadcaster an individual company that owns a number of broadcast stations. The common ownership offers economies in purchasing programs or equipment and in sharing staff and other expenses. Group ownership also can be used as a marketing tool in which would-be advertisers are offered discounts on COMMERCIAL TIME if they "buy the group." (See also CROSS-OWNERSHIP RULE, DUOPOLY RULES, and TRAFFICKING RULES.)

group buy the process by which a collection of PUBLIC TELEVISION (PTV) stations temporarily band together to purchase the license to air a specific program or series of programs. The practice is not often found in the commercial industry, where competition, not cooperation, among stations is the norm. The chronically underfunded public television stations, however, often merge to purchase programming, with each station sharing the cost.

Group W a GROUP BROADCASTER that is a wholly owned subsidiary of the Westinghouse Electric Company. It is officially known as the Westinghouse Broadcasting Company. The company owns 11 AM radio, 12 FM radio, and five TV stations. Long noted for its high standards of broadcasting and attention to news and public affairs, the group has been a major force on the communication scene since the 1960s.

GTE Spacenet Corporation a company that provides a number of services from its KU-BAND and C-BAND Spacenet and G-Star series of communications SATELLITES. Permanent television and cable networks, temporary users and *ad hoc* networks use the company's satellites for business communication and news and sports broadcasting. The firm is the major provider of satellite time and communication packages for SATELLITE NEWS GATHERING (SNG) in the United States.

guerrilla TV See ALTERNATIVE TELEVISION.

H

halo effect the subjective positive reaction to a product or program by a consumer or viewer. Sometimes called the "socially desirable response," it is based on the overall favorable image of the product or type of programming. People are often unable to define any negative aspects or feature of a product because their overall impression of it is so positive, largely due to its excellent reputation.

Consumers wish to identify with quality. Because of the halo effect, audience research firms such as A.C. NIELSEN rely on their DIARY SYSTEMS rather than the statements of viewers.

hammock a program scheduling technique by which a weak show is scheduled between two strong ones on a local television station, network, or cable operation. The strategy is utilized to create an audience for a new program or to build an audience for a show that is faltering in the RATINGS. The technique relies on an AUDIENCE FLOW from the previous program, and usually requires a very strong LEAD-IN PROGRAM in order to be successful.

hangers lighting accessories formally known as "antigravity hangers." They are used to hang lights on a pipe grid above a television studio. The devices are accordion-like mechanisms that are counterbalanced with springs and thus allow the easy raising and lowering of a light to any height. They are used most often to support SCOOP LIGHTS.

hardware the various machines and devices that are manufactured and sold to consumers, such as videocassette and videodisc machines. The term is also sometimes used to describe the physical components of the television and cable industries such as satellite dishes and switchers. In the computer world, it describes the terminals, keyboards, and cathode ray tubes (CRT). (See also SOFTWARE.)

hash mark a mark (—or <<) that sometimes appears across from a program or station in rating books. It indicates that audience NUMBERS were too low to report. The use of the mark does not mean that there was no audience, but instead indicates that it was not large enough to meet minimum reporting standards (MRS) as defined by the particular research company.

HDTV systems See SIMULCAST HIGH DEFINITION TELEVISION (HDTV).

headend the control center for a cable system. The headend is an electronic nerve center, filled with hundreds of pieces of technical gear, and is the origination point for all of the signals transported on a cable system. (See also TRUNK LINES.)

Health Sciences Communication Association (HeSCA) a nonprofit organization that brings together people who are involved in BIOMEDICAL COMMUNICATIONS. This includes producers and developers of AUDIOVISUAL COMMUNICATIONS media and materials, and other professionals involved in using television and video for instruction and information in the health care industry.

Hearst Corporation a major publisher, GROUP BROADCASTER, and cable MULTIPLE SYSTEM OPERATOR (MSO). The corporation publishes a number of newspapers as well as many consumer magazines. It also publishes trade magazines in various fields and owns Avon Books, Arbor House Publishing Company, and William Morrow and Company. The company also owns and operates cable systems in California and has interests in three BASIC CABLE networks, ARTS AND ENTERTAINMENT (A&E), LIFETIME, and ESPN, INC. As a group broadcaster, the company owns four AM and three FM radio stations and six television stations.

heavy-up advertising the practice of scheduling a number of SPOT ANNOUNCEMENTS during a specific short period of time on a television or cable operation. The COMMERCIALS are run in a concentrated manner during a period when the products being promoted are likely to be used the most, such as toy commercials at Christmastime.

helical-scan videotape recording a method that has become the standard for professional and in-home recording throughout the world. The tape wraps itself around a stationary drum containing the rotating recording heads in a helix (or spiral) manner. This diagonal movement of the tape across the horizontally moving heads creates a slanted track on the tape; thus, the term "slant-track," which is often used to describe helical-scan machines. This recording angle enables more information to be recorded on narrow tape.

herringbone effect a pattern of interference in a television picture. It consists of a stationary row (or rows) of saw-tooth images that appear in a diagonal or horizontal manner across the screen. The effect is often caused by a lack of synchronization (SYNC) in the signal and in its extreme can all but obliterate the picture.

Hertz (Hz) a unit of FREQUENCY that, by international agreement, is one cycle per second. The term is named for the German physicist Heinrich Hertz, whose verification of a theory in 1898 ultimately led to the invention of radio. The term is often abbreviated Hz. To represent higher frequencies, metric prefixes such as KILO (for thousand), MEGA (for million), and GIGA (for billion) are commonly added as a prefix to the basic Hertz unit.

hidden camera the placing of a camera in a concealed spot, which often captures the spontaneous and genuine reactions of ordinary people. The unrehearsed bits are used in COMMERCIALS and on programs. Hidden cameras are used intentionally to record blunders as well as to observe everyday occurrences.

high definition television (HDTV) See SIMULCAST HIGH DEFINITION TELEVISION (HDTV).

HMI light a type of light and lighting instrument used in both film and television production when a great deal of illumination is needed. The initials stand for halogen metal iodide, the material that constitutes the light bulb used in the appliance. It uses low power and generates little heat, but creates a great deal of light. (See also COOL LIGHT.)

Home Box Office (HBO) the oldest national PAY (PREMIUM) CABLE SERVICE network in the United States. It is credited with the creation of the modern cable industry. In 1975 HBO revolutionized the industry by linking SATELLITES to cable. Owned by the megaglomerate TIME WARNER, the HBO service offers first-run motion pictures and SPECIALS including a number of boxing and comedy shows.

Home Shopping Networks I and II (HSN) two home shopping networks that are carried by television broadcast and cable affiliates and also air on HSN's 11 O & O (owned-and-operated) television stations. Merchandise, ranging from jewelry and clothing to collectibles and electronic items, is offered at discount prices. A club atmosphere, show-biz hosts, and celebrity guests add an entertainment element to the "mall-in-the-living-room" shows.

homes passed the number of living units (homes, apartments, hotels, motels, condominiums, or other MULTIPLE DWELLING UNITS) that are able to receive cable services. A home is deemed to be "passed" if it can be connected to a FEEDER CABLE or TRUNK LINE by connecting a CABLE DROP to the housing unit. All homes passed do not necessarily subscribe to cable services, but the number of homes that can be hooked up indicates the potential subscribers in an area and in the aggregate on a national basis.

homes using television See HOUSEHOLDS USING TELEVISION (HUT)

hot switch a program scheduling technique in which there are no STATION BREAKS between programs. Sometimes called a "seamless transition," the strategy is used to try to keep viewers from GRAZING and to increase AUDIENCE FLOW. It results in more COMMERCIALS within the programs.

house drop See CABLE DROP.

households using television (HUT) the percentage of households in a DESIGNATED MARKET AREA (DMA) that are actually tuned into stations in that market in a given time period. "Households" are defined as any type of housing unit, including apartments, single rooms, and houses. The concept is sometimes called "homes using television."

HUT measurements determine the level of viewing in the market as a whole. If there are 10,000 television households in the DMA and 4,000 are watching TV at a given time, the HUT level for that market at that time is 40 percent. When this method is applied to people, it is called PEOPLE USING TELEVISION (PUT). SHARES (rather than RATINGS) are derived from HUT levels.

how-tos films, programs, and videos of an instructional nature. Some are series, but many are single programs. There are at least 23 recognized subject categories in the how-to field in home video. They are Animals and Pets, Art, Auto Repair and Maintenance, Aviation, Boating, Child Care and Parenting, Computers and Computer Technology, Cooking, Exercise and Fitness, Fashion and Beauty, Foreign Language Instruction, Gardening, General Health Care, Hobbies and Crafts, Home Repair and Improvement, Legal, Mathematics, Money Management, Music and Musical Instruments, Personal Growth, Sex, and Sports and Recreation. The greatest number of titles in the GENRE are in sports and recreation.

hub system a cable system in which several sub-HEADENDS are used throughout the distribution network to reach subscribers. Such a system is necessary in large cable operations because of the gradual degradation of the signal carried through the COAXIAL CABLES of a cable system. More than one headend is therefore necessary to serve a larger FRANCHISE area. Each of the subordinate centers serves as the hub for some five miles around that

headend. Hub cable systems are typically used in urban settings rather than the more conventional TREE NETWORKS.

Hubbard Broadcasting Inc. (HBI) a GROUP BROADCASTER, a SATELLITE NEWS GATHERING (SNG) organization, and a fledgling DIRECT BROADCAST SATELLITE (DBS) provider. As a group broadcaster, the company owns an AM and an FM radio station and nine television stations in Florida, Minnesota, and New Mexico. HBI also owns a television production firm, Hubcom (an electronic supply firm that creates SATELLITE NEWS GATHERING [SNG] vans), and a hotel. It is the founder and general partner in the national SNG company CONUS COMMUNICATIONS and is one of the prime proponents of DIRECT BROADCAST SATELLITE (DBS) with its UNITED STATES SATELLITE BROADCASTING (USSB) company.

Hughes Communications a company that provides a number of KU-BAND and C-BAND communications SATELLITE services from the Galaxy, SBS, and WESTAR series of satellites. Many television stations and networks and cable operations use the transponders on Hughes satellites for interactive data, voice, SATELLITE NEWS GATHERING (SNG), and video services.

hypermedia a theoretical, futuristic, electronic system that will manage the storage and retrieval of all human knowledge and make it available to every person on the planet in a massive relational data base. The term was coined by Ted Nelson in the late 1980s.

The prefix *hyper* means "over" or "beyond," and *media* alludes to "channels of communications." The term describes a sophisticated system of technology that combines all elements of written and audiovisual communications into an integrated data base that can be accessible via a personal computer (PC). Film, video, computer graphics, still pictures, music, voice, and text will all be combined in an information-delivery system. Any piece of data (called a node) in the world can be connected to another in a logical manner and the system user will have a sophisticated guide to allow the navigation and retrieval of the information.

hyping a short-term practice of scheduling high-quality or unusual programs during a SWEEPS period to influence the RATINGS. Hyping is designed to temporarily inflate the number of viewers of the programs or stations during the sweeps. Sometimes spelled and pronounced "hypoing," the practice is so common that the audience research companies often acknowledge it with "stickered reports." A special sticker appears on the cover of the BOOKS to indicate that hyping has occurred by one or more stations in the market during the survey and that this may have affected the ratings.

Hz   See HERTZ.

I

iconoscope tube an early television pickup tube invented by Vladimir K. Zworykin. Some 13 inches long with a large electron gun at the rear, it required an enormous amount of light to create an image. Zworykin named the tube after the Greek words *eikon* (image) and *skopein* (to view). It had its first public demonstration in 1928 and was used until after World War II, when it was replaced by the more sensitive and smaller IMAGE ORTHICON TUBE. Because of his invention of the iconoscope camera tube and the KINESCOPE TUBE, which displayed the

image, Zworykin is often called the "father of television."

image commercials   COMMERCIALS that are designed to create a general perception of a company and/or a product, rather than to highlight particular aspects of the firm or its merchandise. Image advertising is usually lifestyle-related and normally does not have a specific sales pitch or call to action.

image orthicon tube a camera CATHODE RAY pickup tube, introduced in 1945. The new tube produced higher-quality pictures than the older ICONOSCOPE TUBE and it was smaller, lighter, and more stable. Most important, it required considerably less light to create good images. The importance of the tube within the industry was reflected in the naming of the Immy (later EMMY) awards in 1949, the ACADEMY OF TELEVISION ARTS AND SCIENCES (ATAS) awards for excellence in all aspects of the field.

The long (15 or 20 inches) cylindrical tube with a three-inch or four-and-a-half-inch face was the workhorse of television for more than 25 years, until it was gradually replaced by the even smaller and more powerful VIDICON, PLUMBICON, and SATICON TUBES and eventually by CHARGE-COUPLED DEVICES (CCD).

impeachment hearings the 1974 congressional hearings concerning the possible impeachment of President Richard M. Nixon, which made compelling television. The hearings were conducted by the House Judiciary Committee and began on May 9, 1974, but only a portion of the opening session was allowed to be televised. For nearly three months the meetings were closed to the public, but the committee finally permitted television coverage to resume. The final six days of deliberations were watched by millions of people. On July 30 the committee voted three articles of impeachment against President Nixon in a dramatic on-camera session. On August 8, Nixon became the nation's only chief executive to resign the presidency.

in-school programming See INSTRUCTIONAL TELEVISION (ITV).

Independent Broadcasting Authority (IBA) a nonprofit corporation responsible for commercial television in the United Kingdom. The organization is sometimes compared to the FEDERAL COMMUNICATIONS COMMISSION (FCC) in the United States and the CANADIAN RADIO–TELEVISION AND TELECOMMUNICATIONS COMMISSION, but it performs a different function in that it actually owns television stations.

Formed to establish competition for the BRITISH BROADCASTING CORPORATION (BBC), the organization supervises 15 independently owned and operated television companies. They produce programs for broadcast through television stations, which they lease from the IBA, and sell COMMERCIAL TIME on their programs. The 15 companies often join together to provide a national service program in PRIME TIME.

independent production companies privately owned companies that conceive, develop, and produce television programs for transmission by cable systems and broadcast NETWORKS and for SYNDICATION. They were so named because they were independent of the major Hollywood film studios that dominated television production in the early days of the industry. Today the independents far outnumber the studios and create most of the programming for the medium. They also develop the majority of programs for home video and for many of the AUDIOVISUAL COM-

MUNICATIONS systems in education, health, or government. Home video WHOLESALERS usually buy their NON-THEATRICAL PROGRAMS from such independent PROGRAM SUPPLIERS.

independent stations   television stations that are not affiliated with or owned by a NETWORK (O & Os). They are often called "indies." Some 20 percent of the television stations in the United States are so labeled. They are represented nationally by the ASSOCIATION OF INDEPENDENT TELEVISION STATIONS (INTV).

industrial television   See CORPORATE TELEVISION.

infomercials   program-length commercials. Often masked as adult talk shows or news SPECIALS, they are actually 30-minute or one-hour commercials for a particular product or service. The manufacturer buys the time from a cable system or station and provides the "program," which comes complete with a host, studio audience, and endless hyperbole. The possibility of deception within and by the programs has led the FEDERAL COMMUNICATIONS COMMISSION (FCC) and the FEDERAL TRADE COMMISSION (FTC) to file actions against some of the sponsors. FCC rules require that infomercials be identified as such at the beginning or end of the program.

Institute of Electrical and Electronics Engineers (IEEE)   the world's largest engineering society. The "I-Triple E" is an international organization with members in more than 130 countries. Its purpose is technical, professional, and societal.

instructional design   a process used in schools and colleges as well as in business, health care, and government to improve the efficiency and effectiveness of instruction, often with the aid of AUDIOVISUAL COMMUNICATIONS devices. The effort is usually a collaboration involving a team of teachers and experts in the field who originate or rework training lessons, individual lectures, units, modules, and entire courses.

instructional media   See AUDIOVISUAL COMMUNICATIONS.

instructional technology   See AUDIOVISUAL COMMUNICATIONS.

instructional television (ITV)   the television systems and programs that are used in a systematic manner in a formal educational environment. The term ITV implies a more specific use of the medium for teaching purposes than the broader and nearly obsolete phrase EDUCATIONAL TELEVISION (ETV) in the United States. When instructional television is used in K–12 grade levels, it is often called "in-school" or "classroom" television, and the programs are sometimes called "telelessons." Instructional television programs for college students and adults are usually called TELECOURSES.

instructional television fixed service (ITFS)   a television transmission and reception system that was established by the FEDERAL COMMU-

NICATIONS COMMISSION (FCC) in 1963 expressly for nonprofit educational use. It operates in the same 2-GIGA-HERTZ (GHZ) microwave frequency band as the OPERATIONS FIXED SERVICE (OFS) and MULTICHANNEL MULTIPOINT DISTRIBUTION SERVICE (MMDS) systems.

ITFS transmission systems operate by broadcasting a 10-watt (or with a waiver, a 100-watt) signal in an omnidirectional manner for a radius of some 25 miles.

Educators use the ITFS systems to broadcast ITV programs to schools within a district and universities use ITFS frequencies to transmit continuing education courses to nearby business establishments or to other locations. ITFS frequencies have been underutilized by educators and the Commission reallocated eight of the ITFS channels to MMDS use in 1983.

Intelsat   See COMMUNICATIONS SATELLITE CORPORATION (COMSAT) and INTERNATIONAL TELECOMMUNICATIONS SATELLITE ORGANIZATION (INTELSAT).

interactive multimedia   a sophisticated electronic system through which an individual can retrieve data from various MEDIA. Based on the MULTIMEDIA concept but controlled by the user at a computer workstation, it is also often referred to as INTERACTIVE VIDEO. The information retrieved is usually contained in a single combined form such as CD-ROM or DVI disc. The form differs from INTERACTIVE TELEVISION in that the control of information received

rests with the individual and usually no broadcast, cable, or satellite transmission is involved in the process.

Interactive Multimedia Association (IMA)   a nonprofit international trade association representing the INTERACTIVE MULTIMEDIA industry. This group includes applications developers, suppliers of hardware and software, system integrators, publishers, distributors, replicators, educators, and users.

interactive television   a type of television in which the viewers, by use of computer or communications technology, actively participate in the program. The term entered the vocabulary of the communications industry in the 1970s and is applied to any number of schemes, services, and devices that feature two-way interaction, most employing a computer element. The various schemes use different technologies, including the telephone, cable, broadcast, and satellite.

While the term *interactive TV* is sometimes used in referring to the simple exchange of information, it is often a part of an INSTRUCTIONAL TELEVISION (ITV) operation that involves the teaching (usually formal and systematic) of secondary and college students and adults. The term is also used to broadly describe the more informal exchange of information in TELECONFERENCING or video-conference situations.

In education, interactive television differs from the more specific INTERACTIVE VIDEO and INTERACTIVE MULTIMEDIA process in that it is usually transmitted via broadcasting, cable television, or SATELLITE to large numbers of viewers in a DISTANCE EDUCATION situation and there is proportionately less control of the process by the individual viewer. Often TELETEXT and VIDEOTEXT technology is used to facilitate the interchange. In 1990, the FEDERAL COMMUNICATIONS COMMISSION (FCC) authorized experiments for PCS services that could lead to two-way video operations, and the next year proposed the establishment of an Interactive Video and Data Service (IVDS) using satellites. Eventually, a HYPERMEDIA system of interfacing with data bases throughout the world may be the ultimate interactive television configuration.

interactive video   a video that involves the active participation of the individual in the communication process. Interactive video can be as simple as coding frames on a prerecorded videocassette so the user can fast-forward to any segment. A more sophisticated form is LASER VIDEODISC (LV) technology, which enables random and instantaneous recall of any segment or frame on the disc. The most advanced interactive technique uses the personal computer (PC) to access and retrieve text and graphs, stills, animation, and motion pictures in different combinations from their storage on a videodisc in an INTERACTIVE MULTIMEDIA configuration. Many in the computer industry use that term in place of interactive video to imply a broader, more inclusive application of various computer-controlled media.

Interactive video (or interactive multimedia) makes use of various types of videodiscs including CD-ROM, CDI, and DVI, separately or in conjunction with a personal computer.

Interactive Video Industry Association (IVIA)   See INTERACTIVE MULTIMEDIA ASSOCIATION (IMA).

Interagency Group for Interactive Training Technologies (IGITT)   an organization of employees of the federal government who are involved in training activities using new media. The IGITT serves as a clearinghouse for the exchange of information about new technologies including INTERACTIVE VIDEO, DISTANCE EDUCATION, and INTERACTIVE MULTIMEDIA.

interconnects   regional organizations that sell COMMERCIAL TIME for a number of systems in a geographic area. Most of the interconnects represent cable systems in the nation's largest MARKETS such as New York, Los Angeles, or Chicago. In selling such time the firm becomes the equivalent of STATION REPRESENTATIVES (REPS).

International Advertising Association (IAA)   a nonprofit association consisting of individuals engaged in advertising in countries outside of the United States. The members come from more than 70 nations.

International Alliance of Theatrical Stage Employees (IATSE)   one of the oldest trade unions in the communications field, founded in 1892. IATSE (pronounced "YAHT-*see*") membership consists largely of people involved in the film industry, but a significant number are employed in television operations at both the network and station level. IATSE members include GRIPS, costumers, videotape editors, publicists, set designers, art directors, and make-up and hair stylists.

International Association of Business Communicators (IABC) a nonprofit organization composed of professionals in public relations and corporate communications. Its member writers, editors, and AUDIOVISUAL COMMUNICATIONS specialists utilize many different types of media to reach a variety of audiences in the business world.

International Association of Satellite Users and Suppliers (IASUS) a membership organization consisting of satellite manufacturers and distributors as well as users. Its purpose is to keep its members informed about developments in satellite communications.

International Brotherhood of Electrical Workers (IBEW) a trade union that represents engineering personnel at television stations including supervisory, maintenance, and transmitter engineers, and videotape editors. It represents employees at CBS but not those at NBC and ABC, who are represented by the NATIONAL ASSOCIATION OF BROADCAST EMPLOYEES AND TECHNICIANS (NABET).

International Catholic Association for Radio and Television *See* UNDA—USA.

International Communications Industry Association (ICIA) a nonprofit membership organization that includes audiovisual and video hardware and software producers and manufacturers, retail store owners, sales representatives, and others involved in AUDIOVISUAL COMMUNICATIONS. The association represents its members before Congress, issues reports and studies, publishes directories and newsletters, and conducts a certification program for professionals in the field.

International Council for Educational Media (ICEM) a nonprofit organization involved in the exchange of information among professionals in AUDIOVISUAL COMMUNICATIONS. It encourages COPRODUCTIONS and information exchange among nations.

International Industrial Television Association *See* INTERNATIONAL TELEVISION ASSOCIATION (ITVA).

International Institute of Communications (IIC) a nonprofit organization that aims to further international cooperation in communications and to encourage and distribute studies about the influence of the media on society. The membership consists of corporate, institutional, and national associations as well as individuals from 71 countries.

International Radio and Television Society (IRTS) a New York-based nonprofit organization that consists of individuals in the communications industry, including professionals as well as students. It holds periodic workshops and sponsors Newsmaker Luncheons in which industry notables address their colleagues on communication issues of the day.

International Society of Certified Electronics Technicians (ISCET) a society that sponsors testing programs for the certification of electronics technicians in audio, video, consumer electronics, and radio and television. Members of the society are technicians who have been certified by the organization.

International Society of Communications Specialists (ISCS) a nonprofit organization of people involved in audio and videotape production. ISCS holds seminars and promotes the use of tape in production.

International Society of Videographers (ISV) See AMERICAN SOCIETY OF TV CAMERAMEN (ASTVC).

International Tape/Disc Association (ITA) See ITA.

International Telecommunications Satellite Organization (Intelsat) a consortium of 119 nations involved in communication SATELLITES. The members are national governments that adhere to international telecommunications agreements. The U.S. representative is the COMMUNICATIONS SATELLITE CORPORATION (COM-

SAT), which managed Intelsat in its early years. Intelsat now operates more satellites than any other organization in the world. All member countries participate in the profits or losses of the satellite ventures.

International Telecommunications Union (ITU) a group composed of telecommunications administrations of participating nations, operating under the United Nations banner. The ITU determines the allocation of radio and television spectrum space worldwide and establishes standards for the telegraph and the telephone. Its decisions about radio, television, and SATELLITE applications and the emerging technologies such as ADVANCED TELEVISION (ATV) at periodic World Administrative Radio Conferences (WARC) are vital for the global management of communications.

International Teleconferencing Association (ITCA) a nonprofit membership association that encourages and promotes the growth of TELECONFERENCING. It provides a clearinghouse for the exchange of information among users, researchers, and suppliers in the field.

International Teleproduction Society (ITS) an organization that promotes and encourages the use of videotape as a medium of communication. The society is an international network of companies and individuals involved in videotape production and POSTPRODUCTION.

International Television Association (ITVA) a nonprofit organization of companies and individuals involved in CORPORATE TELEVISION. Its members are in charge of television and media operations at banks, insurance companies, hospitals, and industrial manufacturers. Many members are small INDEPENDENT PRODUCTION COMPANIES or individuals who freelance in the field.

International University Consortium (IUC) an organization of colleges and universities that develops and distributes TELECOURSES designed for DISTANCE EDUCATION use by its member institutions. Located at the University of Maryland, IUC makes courses available to members and nonmembers, and hosts semiannual conferences in the spring and fall.

Interregional Program Service (IPS) an organization of 168 of the nation's PUBLIC TELEVISION (PTV) licensees. It provides a wide range of programming and program services to its membership throughout the nation. IPS was formed in 1980 by the trustees of the EASTERN EDUCATIONAL NETWORK (EEN), working in cooperation with the three other REGIONAL TELEVISION NETWORKS. It operates a GROUP BUY and program SYNDICATION service similar to those of the regional networks.

interstitional programming See FILLER PROGRAMMING.

inventory There are three applications of this term in the communications field. In the advertising industry, it refers to the total amount of COMMERCIAL TIME and SPOTS that are available for sale by a television station, network, or cable operation. Stations, networks, cable systems, and STATION REPRESENTATIVES (REPS) keep a running account of their inventory, in order to accommodate advertisers or ADVERTISING AGENCIES that seek AVAILABILITIES for the time periods.

In broadcasting and cable, inventory is the amount of programming that has been prepared or licensed but not yet broadcast. Stations buy FILM PACKAGES and syndicated programs for transmission over a period of time, and the motion pictures or program titles that have not yet been broadcast are said to be the inventory.

In home video and CONSUMER ELECTRONICS, the term refers to the total number of prerecorded videocassettes or electronic devices such as videocassette machines that are in stock in a retail store at any given time. The retailer must keep a reasonable (but not excessive) amount of inventory to be able to serve customers.

iris an adjustable circular opening that controls the amount of light entering a television, film, or still camera lens. Made of metal or plastic, the iris is adjusted by a ring located on the barrel of the lens. When the ring is rotated, interlaced flaps are closed or opened in varying degrees, permitting different amounts of light to enter the camera through the lens. The size of the iris opening is measured in F-STOPS.

ISDN   an integrated services digital network in development by the AMERICAN TELEPHONE AND TELEGRAPH COMPANY (AT&T) and others as the information highway of the future. Often called the "smart service," it will dramatically increase the origination and distribution of all electronic signals using FIBER OPTIC cable.

isolated camera   a camera that is separate from the others covering an event, which concentrates on a single individual or specific action within the event. Called an "iso," the segregated camera is often attached to its own videotape machine. The technique has become standard for most sports coverage and has also been used in documentaries and occasionally on news programs.

ITA   officially the International Association of Magnetic and Optical Media Manufacturers and Related Industries. The nonprofit organization gathers and disseminates sales statistics on such products as blank audio and videotape and hosts a variety of technical and marketing-oriented seminars that cover the audio, video, and data-storage industries.

J

Jesuits in Communication in the U.S. (JESCOM)   an organization that brings information about communications to U.S. members of the Society of Jesus (Jesuits) of the Roman Catholic church. JESCOM disseminates information about writing, directing, and producing radio and television programs.

Jewish Television Network   a BASIC CABLE network offering news, religious and arts programming that pertains to the Jewish community.

jitter   a technical aberration in a television picture that consists of a jerky unstable jumping of the image. The rapid unsteady effect is often caused by improper synchronization (SYNC) in the playback of a videotape recording.

Jumbotron   a giant presentation television unit that measures 23-1/2 feet by 32 feet and is used for outdoor display. Developed by the SONY CORPORATION, the Jumbotron is used to screen travelogues, commercials, PUBLIC SERVICE ANNOUNCEMENTS (PSAs), videos, and headline news to thousands of passersby from buildings in New York's Times Square and in Tokyo and other cities.

jump cut   an undesirable CUT in television and film production that occurs when a director or editor makes an abrupt transition from one camera angle or scene to a similar one for no apparent reason. A jump cut is also created by an extreme change between shots. In both cases the effect is to make it look like the picture has "jumped" from one shot to another.

K

Kelvin (K) a unit of measurement used in photography and broadcasting to calibrate the COLOR TEMPERATURE of a light source. It is sometimes referred to as degrees Kelvin or °K or simply K. In the Kelvin temperature scale, $0°$ K is equal to $-273.15°$ C.

key a sophisticated form of SUPERIMPOSITION made possible by the introduction of SPECIAL-EFFECTS GENERATORS (SEG) in television production. They key effect replaces part of an image with a different image.

Keying is most often used with graphics to insert words, phrases, logos, or names into other pictures. The resulting definition is crisp and clean and far superior to the similar effect achieved by superimposition. For maximum readability, many directors use an even more sophisticated version of a key called a "matte," which permits the color and brightness of the inserted words to be manipulated to make even cleaner and crisper lettering.

key lighting a television and film lighting technique that directs light onto a performer, object, or scene from the top and side. Key lights are highly directional and are the primary and most concentrated lights used to highlight the subject in a dramatic way by creating shadows, often with the aid of BARN DOORS. Along with FILL LIGHTING, BACKGROUND LIGHTING and BACKLIGHTING, key lighting is one of the four basic illumination techniques in film and television production. (See also SPOTLIGHTS.)

keystone an undesirable visual effect created when a camera is not positioned precisely perpendicular to the plane of a television graphic. It is the result of a camera shot that is just off-center or at an angle from the artwork, with the result that the near side of the graphic appears somewhat larger than the far side. Keystoning is most noticeable with lettering. When the camera is not at the correct angle in relationship to the lettered card, the lettering appears distorted. The term is derived from the trapezoidal or "keystone" geometric shape.

kicker See BACKLIGHTING.

kid fringe a DAYPART in a television station or cable operation that is generally considered the period between 3:00 P.M. and 5:00 P.M. eastern standard time (EST) Monday through Friday. Although the period varies from station to station, the time slot is most often filled with children's shows and cartoons on INDEPENDENT STATIONS. It immediately precedes the EARLY FRINGE period.

kilohertz (KHz) a unit of FREQUENCY in the electromagnetic spectrum designating 1,000 cycles (vibrations) per second. (See HERTZ [HZ].)

kinescope recording   the first method of preserving a television program on film. Introduced in 1947, the technique used a 16mm film camera to record images and sound from a CATHODE RAY picture (KINESCOPE) TUBE. The camera was synchronized with the television image and focused on an especially bright MONITOR. A program was usually recorded from beginning to end and the negative was developed as in regular film processing. The resulting product was called a "kine" (pronounced "kinney"). It was often grainy, of poor contrast, and inferior to regular film quality. The combination electronic/film technique was, however, the only means of recording at the time, and the networks made kines for archival purposes and to supply programs to noninterconnected stations. Kinescopes were quickly abandoned as a means of recording in 1956, with the introduction of VIDEOTAPE RECORDING.

kinescope tube the tube that made electronic television possible. It was invented by Vladimir K. Zworykin in 1925. In combination with the camera pickup ICONOSCOPE TUBE (patented by Zworykin the year before), the kinescope picture receiving tube completed the essential linkup between transmission and reception. The kinescope tube is a CATHODE RAY vacuum tube that operates by shooting a beam of electrons onto the interior of the face of the tube, which is coated with phosphorescence. When the electrons hit the face of the tube, the phosphorescence glows. The kinescope tube converts the video signal from the pickup tube to an identical pattern on its face. Although there have been many receiver improvements with the use of solid-state electronics over the years and three electron beams (for color) have replaced the single monochromatic tube of Zworykin's day, the principle remains the same.

kitchen debate an informal televised exchange that took place in Moscow in July 1959 between Vice President Richard M. Nixon and Soviet Premier Nikita Krushchev. At the time Nixon was touring an international exhibition of goods and equipment with Krushchev. The two Cold War antagonists agreed that their comments could be televised. As they moved through the hall, they exchanged guarded insults and discussed each country's politics, particularly at an exhibit of U.S. kitchen appliances. When the sequence was televised in the United States it appeared as if the vice president had stood his ground against the Communist dictator and the event enhanced Nixon's politically rising star.

Kodak still-picture process a hybrid technique of traditional and electronic photography from the Kodak Company. Jointly developed with the Philips Corporation, it uses the traditional photographic method but also permits pictures to be manipulated and viewed electronically. The new system is called "Photo CD."

A consumer receives 35mm negatives and prints from the photofinisher in the traditional manner. However, the images can also be converted to digital data and recorded on a COMPACT DISC (CD) and can then be played back on a televi-

sion screen by using a special player. Each disc can contain as many as 100 pictures from either slides or negatives.

Ku-band satellites communication SATELLITES that contain TRANSPONDERS on FREQUENCIES in the KU-band from 11.7 to 12.2 GIGAHERTZ (GHZ). Signals of these high frequencies can be received by relatively small TELEVISION RECEIVE ONLY (TVRO) dishes, thus making DIRECT BROADCAST SATELLITE (DBS) services possible. Ku-band transmission is sensitive to atmospheric changes, however, and satellites utilizing that BAND WIDTH cover only a small portion of the United States, in contrast to C-BAND satellites, which cover most of the country.

L

label an identifying word or phrase, often a brand name, attached to a consumer electronic device or prerecorded videocassette or videodisc. A standard term used in the consumer product industry, label was adopted by the music recording industry in its early days and later picked up by home video companies. A label is usually registered under the Lanham Trademark Act of 1946.

Lar Daly Amendment a 1959 amendment to Section 315 of the

COMMUNICATIONS ACT OF 1934, which addressed the issue of the EQUAL TIME (OPPORTUNITY) RULES for political candidates in broadcasting. It was named after a perpetual candidate for mayor in Chicago. Although he had absolutely no chance of victory, Daly claimed equal access to television time whenever the incumbent mayor appeared on a newscast. The FEDERAL COMMUNICATIONS COMMISSION (FCC) ruled that the maverick was entitled to equal access. Congress then rushed through an amendment to the Act, which ruled that *bona fide* newscasts, documentaries, interviews, and on-the-spot coverage of news events were exempted from equal time (opportunity) considerations.

laser a device for generating a beam of coherent electromagnetic signals. It produces a highly concentrated beam of light at a single FREQUENCY or color, and this beam can then be transmitted in a variety of ways. The word laser is an acronym for "light amplification by stimulated emission of radiation." Although the technology has a number of uses in medicine, education, and business, its first application in the mass media was in the development of COMPACT DISCS (CD) and LASER VIDEO DISCS. Lasers are also utilized in cable television in combination with FIBER OPTICS technology.

laser videodisc (LV) electronic optical machines introduced in 1978. Although the first machines could not record, the laser videodiscs had a staggering capacity for the storage and playback of prerecorded material. More than 54,000 still-frame pictures could be encoded on one side of a disc, or one 30-minute side could play back the graphics and text of 30 books. The machine contains low-power LASERS that reflect off the surface of the videodisc and create electronic signals that can be seen on the screen when the device is attached to a TV set.

The 12-inch discs are plastic coated, making them resistant to dust or scratching, and they will supposedly never wear out because only the laser beam makes contact with the disc. Depending on how the information is mastered, the LV machine will scan at a certain number of revolutions per minute and play back in the CAV (constant angular velocity) or CLV (constant linear velocity) modes. The CAV mode has a program capacity of 30 minutes per side and the CLV mode increases the playback capability to one hour per side. The machine languished in the marketplace until the introduction of the so-called "combi-players" in the late 1980s. The consumer was then offered an affordable laser machine that could play both videodiscs and COMPACT DISCS (CD).

In 1990 Panasonic began selling the first commercially available "rewriteable" optical disc recorder. The WORM (**write once, read many** times) disc can go through as many as a million erase-and-rewrite cycles. Full-motion video, still images, or other data can be replaced at will on a frame-by-frame basis and information can be refined in as little as 7/10th of a second. (See also CD-I, CD-ROM, DVI, and LASERDISC ASSOCIATION.)

LaserDisc Association a New York-based organization founded by hardware and software companies involved in the LASER VIDEODISC (LV) format. Composed of more than 30 firms including the 13 companies that manufacture or market laserdisc players, the group promotes the format to the trade and consumers.

late fringe a DAYPART that is generally considered to be the time slot between 11:30 P.M. and 1:00 A.M. on affiliated stations, although the period varies from station to station. Because most INDEPENDENT STATIONS do not produce local newscasts, the late-fringe period begins at 11:00 P.M. on those outlets. (All times are eastern standard time [EST] Monday through Friday.)

late night   in broadcasting the time slot beginning from 1:00 A.M. (eastern standard time [EST] Sunday through Saturday) until morning on local television stations, networks, and cable systems. During this DAYPART older movies, syndicated programs, and INFOMERCIALS are generally scheduled.

lavaliere microphone   a small omnidirectional microphone, usually referred to as a "lav." It is very popular for television talk shows and interviews and is worn on a cord around the neck or clipped to a lapel. Modern lavaliere mikes are so small they are nearly invisible and their only limitation is the mike cord length. They are named after Madame de La Vallière, a onetime mistress of Louis XIV, who always wore a jewel suspended on a chain above her bosom.

lead-in program   a show on a television station, network, or cable system that begins a DAYPART or a particular sequence of programming. The programming strategy is to schedule a strong initial program with great appeal in order to create good AUDIENCE FLOW for the remainder of the time period.

leader   a short piece of film or videotape that is placed on the front of a videotape recording or film used in television. The practice was originally developed and encouraged by the SOCIETY OF MOTION PICTURE AND TELEVISION ENGINEERS (SMPTE) for film and today both film and videotape leaders are known more formally as "Society leader" or "SMPTE leader." The film leader consists of some 30 seconds of film that contains 15 or 20 seconds of black blank film (for threading the projector) and 10 seconds of timing numerals. In addition to the timing function, an SMPTE leader on a videotape provides electronic test signals to allow technicians to adjust the settings on the videotape recorder before a full playback or dubbing operation. It is 45 seconds in length.

leading/lagging chrominance effect   a technical aberration in a television picture. This effect occurs when the chrominance (color) portion of the video signal leads or lags behind the luminance (black-and-white brightness) signal. The result is an undesired effect in which the colors appear to the left (leading) or to the right (lagging) of the image.

The Learning Channel (TLC)   a BASIC CABLE service that delivers educational programming, business and career information, and how-to and personal enrichment programs to cable subscribers on a 24-hour basis. The network acquires TELECOURSES from educational distributors (including the AGENCY FOR INSTRUCTIONAL TECHNOLOGY [AIT] and the ANNENBERG/CPB PROJECTS) and produces programs in house. TLC is owned by the DISCOVERY CHANNEL.

Learning Link   an in-service teacher information project developed by PUBLIC TELEVISION (PTV) station WNET in New York. It consists of a nationwide computer network that allows teachers to access a number of information data bases concerning education, in order to research instructional materials and review new products and INSTRUCTIONAL TELEVISION (ITV) PROGRAMS. While the CENTRAL EDUCATIONAL NETWORK (CEN), located in Chicago, assumed fiscal and management functions for the project in 1990, Learning Link staff members operate out of both New York and Chicago. The project complements EDISON, another computer-based service also operated by CEN. Both are supported by grants from THE CORPORATION FOR PUBLIC BROADCASTING (CPB) and user fees.

learning resources   See AUDIOVISUAL COMMUNICATIONS.

leased access channels   See CUPO LEASED ACCESS CHANNELS.

least objectionable program (LOP) a famous, half-serious programming theory postulated in 1971 by NBC program executive Paul Klein. He theorized that most viewers do not tune in to watch a specific program but simply turn on the set to watch television, and they will stay tuned to the least objectionable program (LOP) rather than turn the set off.

letterboxing   a method of showing widescreen motion pictures in their original dimensions on television and, with increasing frequency, on home video. Letterboxing is necessary because most theatrical motion pictures are shot in a format that is incompatible with television. They have an ASPECT RATIO of about 9:16, while television screens are more square, with an aspect ratio of 3:4. In letterboxing, a more rectangular image (usually in a 1:1.85 or a 1:2.35 aspect ratio) is seen on the television screen, with black bands below and above the picture filling in the empty spaces on the screen.

Library and Information Technology Association (LITA)   a division of the American Library Association (ALA) concerned with computers and AUDIOVISUAL COMMUNICATIONS techniques and equipment, including videocassettes and LASER VIDEODISCS (LV). The members of this nonprofit organization are also involved in library automation and cable systems.

Library of Congress: Motion Picture, Broadcast, Recorded Sound Division   the largest archive collection of entertainment and information programming in the world. It is a division of the Library of Congress, housing 100,000 motion picture films, 500,000 radio programs, and 80,000 television programs. The division has 14,000 television programs dating from before 1979 and also permanently stores copies of the VANDERBILT TELEVISION NEWS ARCHIVE video tapes. (See also AMERICAN MUSEUM OF THE MOVING IMAGE [AMMI], ATAS/UCLA TELEVISION ARCHIVES, MUSEUM OF BROADCASTING COMMUNICATIONS [MBC], MUSEUM OF TELEVISION AND RADIO, and NATIONAL CENTER FOR FILM AND VIDEO PRESERVATION.)

license   See FCC LICENSE.

license renewal   the process of extending a broadcast station's authorization to continue operations. The FEDERAL COMMUNICATIONS COMMISSION (FCC) issues licenses to operate broadcasting stations for a specific period of time. The current length is seven years for radio stations and five years for television operations. An FCC LICENSE is also issued for LOW POWER TELEVISION (LPTV), MULTICHANNEL MULTIPOINT DISTRIBUTION SERVICE (MMDS), and INSTRUCTIONAL TELEVISION FIXED SERVICE (ITFS) stations. All licenses may be renewed "if the public convenience and necessity would be served" by a renewal. Unless a license is challenged by another company or group resulting in a COMPARATIVE HEARING, the license is usually renewed. Most licenses (98 percent) are uncontested and are renewed by the staff of the FCC.

Lifetime   a BASIC CABLE network featuring programming for and about women. Lifetime airs programs related to all aspects of women's lives including issues and lifestyles as well

as food and children. It is owned by CAPITAL CITIES/ABC, VIACOM ENTERPRISES, and HEARST.

lift   the process of increasing the number of subscribers at a cable television system. Business can be increased by pursuading BASIC CABLE SERVICE subscribers to purchase PAY (PREMIUM) CABLE SERVICES, and then persuading those subscribers to add PAY-PER-VIEW (PPV) selections. The term also refers to the acquisition of first-time subscribers.

light meter   an electronic instrument that measures the level of light utilized in a television or film production. The hand-held device contains a photoelectric cell that measures light reflected off a subject, ambient light, and direct light. The measurement is expressed in FOOTCANDLES (FC) or lux.

lighting plot   a diagram showing the position of all lighting instruments in the production. The layout is sketched as seen from above and shows the placement of KEY, FILL, and BACKLIGHTS. It is prepared in advance of the show and acts as a blueprint for the positioning of SCOOP LIGHTS and SPOTS, which are often represented by symbols. It sometimes also indicates the DIMMER circuit numbers for each instrument. (See also FLOOR PLAN.)

lighting ratio   the relationship between light and dark in a television picture. The film camera can distinguish something that is 100 times brighter than another part of the picture, but the television camera is less sensitive. It can only handle a difference of approximately 30 times brighter (a 30:1 ratio) than something else in the same scene without creating a blooming or washed-out effect. One way of measuring the lighting ratio in a television set is to determine the brightness level of the FILL LIGHTING compared to the brightness level of the KEY LIGHTING as measured in FOOTCANDLES by a LIGHT METER. The ratio is derived by dividing the lower number of the fill light into the higher number of the key light.

limited series   See MINISERIES.

limited use discount (LUD) policy a cost-saving device used by some PUBLIC BROADCASTING SERVICE (PBS) stations that cannot pay for (or do not want to broadcast) the complete PBS program schedule. A station that does not purchase all of the service is restricted to no more than 50 percent of the program package, only 33 percent of which may come from the PRIME-TIME schedule. Some 22 PBS station licensees (of a total of 192) have historically purchased less than 100 percent of the PBS schedule.

limited-play videocassettes a type of videocassette that can be rented and watched for a limited number of times before automatically erasing itself. It has a built-in counter that notes how many times it has been viewed and an internal magnet that erases the tape after 25 screenings.

The consumer must pay for each viewing in a PAY-PER-VIEW (PPV) type of strategy for home video. When the tape is returned, the counter shows if the tape has been played more than once, and thus if any further charge is due.

lip synchronization a technique used in both television and film production, that matches the voices of performers speaking or singing with their lip movements. It is called "lip sync" for short. The technique was frequently used on Dick Clark's "American Bandstand" show in the 1950s and 1960s when hundreds of rock stars and groups, performing live, lip-synced their big hits (which had been recorded in a sound studio using elaborate instrumental backgrounds). The height of the deception concerning lip syncing occurred in 1990 when it was revealed that the Grammy Award winners Milli Vanilli were not the real singers on their hit records, concerts, and music videos. Lip syncing is often used today in television on musical-variety shows where it allows frenetic stars and dancers to breathlessly perform to prerecorded songs. It is also used to correct flubs spoken by actors in prerecorded dramatic shows.

liquid crystal television   See FLAT-PANEL TELEVISION.

local authority stations a type of PUBLIC TELEVISION (PTV) station licensed by the FEDERAL COMMUNICATIONS COMMISSION (FCC) to a governmental agency in a community. Such operations are the smallest group of PTV stations. In 1993 the nine stations of this type were owned by public school systems.

local origination channels specific cable channels designated for locally originated programming. Some of the programming is as simple as slides or billboards for a weather forecast. Many are "swap channels" listing items for rent or sale or community calendar channels promoting local events. In some locations, the channels are used to carry local broadcasts of high school ball games or holiday parades and other community events, often supported by advertising. A number of cable systems cover city council or school board meetings and many candidates appear on panel discussions and interviews during election years. While PEG and CUPU LEASED ACCESS CHANNELS are also used to transmit local programs, they are distinct from local origination channels because their content is usually controlled by a body other than the cable system.

lockbox a device that allows a cable subscriber to block out reception of a particular channel at any given time. It is installed at the back of a television set and contains a TRAP that can be activated by a key.

Such a box protects those who do not want to receive what they consider to be objectionable, obscene, or indecent programming. The CABLE COMMUNICATIONS POLICY ACT OF 1984 requires a cable operator to provide (by sale or lease) such a box at the request of the subscriber.

long shot (LS) a type of shot from a television or film camera that encompasses the entire scene and involves a wide view of the area. It is sometimes referred to as a "wide" or "full" shot or as a CUTAWAY when it hides errors created by JUMP CUTS. When a long shot is used in the opening of a sequence it is often called an "establishing" shot, in that it orients the audience to the surroundings or the circumstances of what they are going to see. The shot is also known as a "cover" shot. (See also CLOSEUP [CU], COMBINATION SHOT, FRAMING, and MEDIUM SHOT [MS].)

longitudinal videotape recording (LVR) a pioneer reel-to-reel VIDEO-TAPE FORMAT developed and demonstrated by Bing Crosby Enterprises in 1951. It operated on the principle of recording electrical impulses on narrow magnetic tape, which moved rapidly over stationary recording heads. The tape had to move 100 inches per second over the heads, however, and the resulting black-and-white image had poor RESOLUTION and produced JITTER. Additionally, fewer than 16 minutes could be recorded on a reel. The LVR-type of recording/playback was soon bypassed for professional use by the QUADRUPLEX (QUAD) VIDEOTAPE RECORDING system in 1956.

loss leader an item that is advertised and sold at a price that represents a loss of profit for the retailer. The pricing technique is used to draw (or lead) customers into the store with the hope that they will buy other items. Loss leaders in home video stores may be bargains on blank tape or on a hot new "A" TITLE to entice customers to make a SELL-THROUGH purchase of a "B" TITLE or CATALOG PRODUCT.

lottery rules   federal broadcasting regulations relating to games of chance. In 1948 Congress removed passages forbidding lotteries on broadcasting from the COMMUNICA-TIONS ACT OF 1934 and placed them in Section 1304 of the U.S. Criminal Code. That prohibition states that a licensee cannot broadcast "any advertisements or information concerning a lottery," and that any licensee doing so is subject to a fine of $1,000 and/or a year's imprisonment. FEDERAL COMMUNICATIONS COMMISSION (FCC) rules also forbid cable systems from using any origination channels to transmit "any advertisement of or information concerning any lottery, gift enterprises, or similar schemes offering prizes dependent in whole or in part upon lot or chance." In 1975 Congress modified Section 1307(a) of the Communications Act and permitted the broadcast of information or advertisement of a lawful state-operated lottery by stations in that state or adjacent states. The FCC amended its rules to conform to the changes.

A newer 1990 law entitled the Charity Game Advertising Act of 1988 relaxed the prohibitions even further. Stations may now broadcast commercials about lotteries authorized by the state in which they are conducted. Advertisements for casinos in Las Vegas or Atlantic City, however, continue to be unlawful.

low power television (LPTV) stations   broadcast stations that transmit a signal to a limited geographic area. They are sometimes called "community broadcasting stations" by their proponents. They utilize the same frequencies as their full-power UHF and VHF brothers but transmit at a lower wattage. They are limited in their power to 10 watts UHF and 1,000 watts VHF and cover about 15 to 20 miles rather than the 50 or 60 miles of a conventional, full-power station. LPTV stations are licensed by the FEDERAL COMMUNICATIONS COMMISSION (FCC) as a "secondary service." As such, if they technically interfere with a primary conventional television station, they must correct the interference or go OFF THE AIR.

LPTV stations are essentially TRANSLATORS that originate programming. The programs can be produced and transmitted locally or brought in from a SATELLITE network via a TELEVISION RECEIVE ONLY (TVRO) dish for simultaneous retransmission. Anyone with a regular television set can pick up the LPTV signal.

The FCC places no restrictions on how the operations are to be supported (by fees, advertising, or donations) nor on the types of programs that an LPTV station can broadcast. There are no CROSS-OWNERSHIP or multiple ownership restrictions, and only limited FAIRNESS DOCTRINE and

EQUAL-TIME (OPPORTUNITY) RULES. The stations, however, have to observe the Criminal Code and LOTTERY RULES as well as the statutory prohibition concerning obscenity. Most operate as a video cross between a local newspaper and an FM radio station.

lowest unit charges (LUC) mandated minimum costs related to political broadcasts. Broadcast stations that sell time must conform to Section 315 of the COMMUNICATIONS ACT OF 1934 (as amended), which outlines the EQUAL TIME (OPPORTUNITY) RULES for political candidates. Part of that section requires a station to charge "the lowest unit charge of the station for the same class and amount of time for the same period . . ." to political candidates. These charges must be applied during the 45 days before a primary election and during the 60 days preceding a general election. The rule ensures that political candidates will be given all the discounts that are offered to the station's most favored commercial advertiser for the same time and period, regardless of how much program time or how many SPOTS the candidate buys from the station. The LUC rates also include discounted rates given to favored commercial advertisers but not published in the station's RATE CARD.

LUD See LIMITED USE DISCOUNT (LUD) POLICY.

lux See FOOTCANDLE.

M

M an abbreviation for one thousand that is used in the ADVERTISING AGENCY world and by audience research firms such as A. C. NIELSEN and ARBITRON. The COST PER THOUSAND for an advertising CAMPAIGN is usually expressed as the CPM. M is the Roman numeral for 1,000, or 10 hundred. (See also MM.)

M format a CAMCORDER recording method that is now considered obsolete. It was used for professional ENG and EFP production. Like the BETA-CAM FORMAT, the two M format types could record for 20 minutes in the field. The units used regular VHS videocassettes but normally required separate playback/editing devices in the studio. The videocassettes could not be played back on regular VHS FORMAT units. (See also COMPONENT VIDEO SYSTEM RECORDING.)

MacBeth color checker a color rendition chart used by film and broadcast engineers to help determine the color accuracy of film and video images. It has become the industry standard for checking color accuracy in film, video, and graphics.

Macrovision a video process involving the encoding of a signal on a videocassette that prevents that cassette from being duplicated properly. It was developed in the early 1980s and introduced to the industry in 1986 to curb PIRACY. Although there are other similar techniques, the Macrovision copyguard has become the *de facto* standard in the video industry in the United States. The encoded signal produces noticeably degraded dubbed copies.

made-for-TV movies sometimes called "telefeatures" or "telepics." The first made-for-TV film to be aired was "See How They Run," broadcast on October 7, 1964. With theatrical film costs skyrocketing and PAY (PREMIUM) CABLE networks competing with broadcast networks for telecast rights, more made-for-TV movies were rushed into production and by the 1978–79 season there were more of them scheduled on the broadcast networks than Hollywood theatrical fare. The TV films were produced at a faster pace and at lower cost. By 1990 two-thirds of the motion pictures on the broadcast networks were specifically made for television. The trend has also affected cable television, where "made-for-pay" features are now produced in abundance.

Madison Avenue a New York City street that is linked closely with the ADVERTISING AGENCY business; it was originally the address of most of the major agencies. Although many agencies have moved elsewhere, the address and its diminutive "Mad Avenue" continue to be synonymous with (and evoke images of) the advertising industry.

magazine format programming a type of nonfiction television FORMAT that is organized into segments, with each section being a self-contained feature. The term is derived from the technique used in laying out print magazines. The format is most often used in news or sports programming and emphasizes features rather than hard news. Three or four brief mini-documentaries within an hour's telecast on a variety of subjects entice an audience. The quintessential magazine format program is "60 Minutes" on CBS.

The standard magazine format show is built around a host or two who introduce the segments from a studio setting and make the transitions from segment to segment. The hosts may also serve as correspondents in the field, doing interviews or STANDUP reports. The segments of some magazine programs are labeled "Act I," "Act II," etc. Some TABLOID TV PROGRAMS such as "A Current Affair" and "Hard Copy" also use a magazine format.

magic hour a time of day, particularly dawn or dusk, that is the ideal period to photograph a scene on a television remote or on a film location. There is little need to adjust the lighting or camera F-STOPS at that time because the COLOR TEMPERA-

TURE is nearly perfect for the conditions of the shoot.

Magnavision See LASER VIDEO-DISC (LV).

makegood the credit that a television station, network, or cable operation must usually give to an advertiser or to its agency for COMMERCIALS that failed to reach the guaranteed number or type of viewers. Makegoods are also given when the commercial did not run because of an error and when the transmission was technically below par. The credit is usually in the form of a rerun or an extra play of the commercial, although all makegoods are negotiable. An alternate term is "bonus SPOTS."

makeready a process carried out by a FRANCHISE (CABLE) holder to make sure that all legal and physical elements are in place before beginning the NEW BUILD of a cable system. The staff of the cable system verifies the location of all poles and confirms that all attachment and easement rights have been cleared. A check is made of all poles to ensure that they will withstand the additional weight of the COAXIAL CABLE, AMPLIFIERS, and other electronic gear. The makeready process also involves a check to be sure that all clearances from the local government and utility companies have been obtained.

margin a term used in the retail industry to indicate the amount of increase included in the sale price of goods over and above their actual cost. It is sometimes called "the markup" and is usually expressed in terms of a percentage. If an item is purchased from the manufacturer for $100 and then is sold for $125, the margin is 25 percent. The size of the markup in retail often depends on the sales volume of the product. Lower margins are taken on items or titles that have a rapid TURNOVER rate.

market a broad and somewhat imprecise term used in the broadcasting and advertising industries to refer to the geopolitical area served by a radio or television station. The area contains a population that buys, sells, and trades in goods and services.

Markets generally conform to the METROPOLITAN STATISTICAL AREAS (MSA) and the CONSOLIDATED METROPOLITAN STATISTICAL AREAS (CMSA) as determined by the federal government and are more precisely defined by the ARBITRON COMPANY with its AREA OF DOMINANT INFLUENCE (ADI) and by A. C. NIELSEN with its DESIGNATED MARKET AREA (DMA). There are more than 200 recognized markets in the United States. (See also ABCD COUNTIES, MARKET-BY-MARKET BUY, and METRO AREA.)

market-by-market buy COMMERCIAL TIME that is purchased in individual markets. It is sometimes referred to as a "national SPOT buy" or simply a "spot buy." In this transaction the time periods are purchased by an ADVERTISING AGENCY for a national advertiser in more than one television market. It differs from a NETWORK BUY in which the advertiser simultaneously purchases local time on all of the stations affiliated with a network. A market-by-market transaction is often less expensive than a network buy if the COMMERCIAL only needs to cover a part of the country. (See also UNWIRED NETWORKS.)

Markle Foundation officially the John & Mary R. Markle Foundation, a philanthropic, nonprofit organization that has had a profound influence on a number of organizations and issues in mass communications. Beginning in 1968 with the startup support of the CHILDREN'S TELEVISION WORKSHOP (CTW), the organization has funded a number of studies on the effects of television on children. The foundation has also supported the programs and projects of the ASPEN INSTITUTE PROGRAM ON COMMUNICATIONS AND SOCIETY, the MEDIA ACCESS PROJECT, the PUBLIC BROADCASTING SERVICE (PBS), and the BRITISH BROADCASTING CORPORATION (BBC).

markup   See MARGIN.

mass media   See MEDIA.

master antenna television (MATV) system   a type of television distribution system designed to receive over-the-air broadcasts in multiple dwellings such as hotels or apartment houses. The internal wiring of MATV systems is often also used to distribute MULTICHANNEL MULTIPOINT DISTRIBUTION SERVICE (MMDS) signals or programs received via a SATELLITE MASTER ANTENNA TELEVISION (SMATV) SYSTEM to residents or guests.

Matsushita   a Japanese company (pronounced mat-SOO-shee-ta) that is one of the world's largest manufacturers of industrial and consumer electric and electronic products. It produces some of the most familiar audio and video gear under the brand names Technics, Quasar, Panasonic, and JVC. It is Japan's largest manufacturer of electronic products and the twelfth largest corporation in the world, with some 117 subsidiary companies in 38 countries.

One of its subsidiaries, JVC, developed the VHS FORMAT in 1976, which prevailed over the archrival SONY CORPORATION and its BETA FORMAT in the VIDEOTAPE FORMAT wars. Today, Matsushita is heavily involved in the development of ADVANCED TELEVISION (ATV) and has also expanded into computers, the half-inch analog COMPONENT VIDEO SYSTEM/RECORDING format (M II), and the half-inch digital COMPOSITE VIDEO SYSTEM/RECORDING format. In a further globalization move, the company purchased MCA INC. in 1990.

matte   See KEY.

MCA Inc.   a diversified international company engaged in the production and distribution of theatrical motion pictures, television, and home video programs. The firm is also involved in the operation of a tour of the company's motion picture studios, the manufacture and distribution of recorded music, music publishing, and the management of amphitheaters. It owns Spencer Gifts, a book publisher (G. P. Putnam's), a television station, and a

50-percent interest in a BASIC CABLE television network. The initials stand for the **M**usic **C**orporation of **A**merica, which was a large talent agency formed in 1924.

McCarthyism the witch-hunt for communists and communist sympathizers in the 1950s that was led by the junior senator from Wisconsin, Joseph R. McCarthy (R). McCarthy began a series of government investigations into alleged communist infiltration in the State Department, government agencies, the military, the Hollywood film community, and broadcasting.

His name became synonymous with the vigilante techniques practiced by self-appointed groups that sought to root out "leftists" and "pinkos," particularly in broadcasting. McCarthy's credibility was badly damaged, however, by his own appearance on "See It Now" in 1954, where his inconsistent logic and rantings alientated viewers. Later that year, during the televised ARMY–MC-CARTHY HEARINGS, he revealed himself to be a posturing demagogue. The Senate finally censured McCarthy for his tactics and irresponsibility and he died a short time later. (See also BLACKLISTING and RED CHANNELS.)

mechanical television the primary technique used in television experimentation until the 1930s. It was first developed in the 1800s.

Based on the principle used in the NIPKOW SCANNING DISC, the technique used a drum or disc that contained a series of perforations in a helical pattern. A special camera was focused on the subject, and the disc was placed between the subject and a light source such as a photocell. When the disc was rotated rapidly, the light shone through each hole, one at a time, and thus "scanned" the object. A lamp at the receiving end varied its brightness depending on the amount of current in the photocell, and the image was seen when a similar disc was rotated and synchronized with the "transmitting" disc, thus reversing the scanning process. The images produced by mechanical television systems usually only contained 30 to 60 SCANNING LINES and were therefore dim and blurred. In the 1930s, a hybrid mechanical-electronic system and later an all-electronic system replaced mechanical television.

media channels of communication, including everything from printed material to electronics. The term is often preceded by an adjective that acts as a qualifying description. Magazines and newspapers are called "print media" while radio and television are called "broadcast media." Cable and home video are often referred to as the "electronic media." AUDIOVISUAL COMMUNICATION devices are commonly referred to as "educational media." Media such as broadcast television that are designed to reach the maximum number of people are called "mass media." A single form of the media is known as a "medium." When more than one medium is used to simultaneously reach an audience, the term "multimedia" is used.

Media Access Project a public-interest law firm in Washington D.C. that seeks to ensure that all media inform the public completely on issues such as civil rights, politics, and the economy. It advises and represents organizations that work to make broadcasting more understanding of their points of view. (See also MARKLE FOUNDATION.)

Media Action Research Center an organization based in New York and founded in 1974 to provide concerned individuals with information to make them aware of television content. It helps individuals conduct Television Awareness training and Growing with Television workshops that study the values of television and compares them with the principles of Christianity.

Media Alliance a nonprofit group composed of writers, editors, broadcast people, and others who support independent journalism. It seeks to discourage competitive practices among people in the field by fostering cooperation and by helping members help one another.

media buy the purchase or the process of purchasing COMMERCIAL TIME for advertising on radio, television, or cable operations or space in the print media for advertising messages. Such purchases are usually made after an advertising CAMPAIGN has been developed by an ADVERTISING AGENCY.

media plan a detailed strategy that lists the specific MEDIA to be used in an advertising CAMPAIGN. Developed by an ADVERTISING AGENCY, the plan will specify the media (such as television, radio, cable, or print) that will be used and the cost and time period for each use. It will also specify the goals of the campaign in terms of FREQUENCY and CUME.

medium See MEDIA.

medium shot (MS) a television shot that covers a portion of the scene or subject but not the entire area. It falls midway between a LONG SHOT (LS) and a CLOSEUP (CU). An MS of a person will usually show the individual from the waist up. (See also FRAMING.)

megahertz (MHz) a unit of FREQUENCY in the electromagnetic spectrum. It is commonly abbreviated MHz and is equivalent to one million cycles (vibrations) per second. This is close to the frequency of video signals. (See HERTZ.)

metro area the geographic core portions of a DESIGNATED MARKET AREA (DMA), as determined by the A. C. NIELSEN COMPANY. Sometimes known as the "central area" or "local DMA," the metro area usually corresponds to the METROPOLITAN STATISTICAL AREA (MSA) as determined by the Office of Management and Budget (OMB) of the federal government. It is the most densely populated part of the MARKET and is important to many local advertisers and ADVERTISING AGENCIES for that reason.

Metropolitan Statistical Area (MSA) an urban territory that is determined by the Office of Management and Budget (OMB) of the U.S. government. Previously called the Standard Metropolitan Statistical Area (SMSA), it usually includes one or more counties (where the population is concentrated) and a central core called the METRO AREA (where the majority of the population is concentrated). The precise dimensions of an MSA change due to population shifts, but usually the MSA must include a city of at least 50,000 people and surrounding communities that are economically, culturally, and socially dependent on the city, with a total population of 100,000. If the area has more than one million inhabitants, it is called a Primary Metropolitan Statistical Area (PMSA). When two or more MSAs are contiguous they are called a CONSOLIDATED METROPOLITAN STATISTICAL AREA (CMSA). (See also MARKET.)

MGM/Pathe Communications Company a major conglomerate engaged in motion picture and pay television production and distribution; television production, licensing, and SYNDICATION; and character and logo licensing and merchandising. One of the major distributors in the industry, the company owns the 1,000-title United Artists library and controls the foreign pay-television rights of the Turner/MGM library of 2,950 pictures until 2001.

The company's origins date back to two of the most famous Hollywood studios, Metro-Goldwyn-Mayer and United Artists. After the two studios merged, the company became MGM/UA. More recently a subsidiary of the TURNER BROADCASTING COMPANY (TBS), the company was purchased in 1986 by the Tricinda Corporation. Turner, however, retained domestic broadcast and cable rights for many of the MGM films and used them as a basis for the formation of the Turner Entertainment Company.

MGM/UA's main library has been the firm's greatest asset and the mainstay of its business, and its films are distributed worldwide in SYNDICATION. In 1990 Pathe Communications Corporation, a European-based entertainment company, purchased the company.

MHz See MEGAHERTZ (MHZ).

microwave relay an electronic system of point-to-point communication. The technology allows for the interconnection of radio, television, and cable systems. Because they operate through the air at high FREQUENCIES, all microwave systems are licensed by the FEDERAL COMMUNICATIONS COMMISSION (FCC).

Using a very short electromagnetic wavelength (generally above 1.6 MEGAHERTZ [MHZ]) focused into a narrow beam, a signal can travel some 30 miles without a great deal of ATTENUATION. In a point-to-point relay system, towers with AMPLIFIERS and small receiving and retransmitting ANTENNAS are set up and the signal passes from tower to tower. The receiving equipment on the tower captures the signal, amplifies it, and retransmits it to the next relay station some 30 miles away. This type of relay system made transcontinental television possible in 1951.

Microwave relay systems are now used to transmit signals from a news site back to the studio or from a studio to a transmitting tower and antenna for rebroadcast. When a microwave relay system is used to connect the studio to the transmitter site, it is called a STUDIO-TRANSMITTER LINK (STL). A microwave relay system used by cable systems to pick up stations that are too far away for OFF-THE-AIR reception is licensed by the FCC as a community antenna relay service (CARS).

midseason See SEASON.

Midwest Video II decision a Supreme Court decision that had an effect on the extent and role of public access channels on cable systems in the 1970s and early 1980s. In its CABLE TELEVISION REPORT AND ORDER OF 1972, the FEDERAL COMMUNICATIONS COMMISSION (FCC) had required cable operators to give access to their facilities to educational institutions, local government spokespersons, and members of the community, without retaining any control over the editorial content of those programs. In a Supreme Court challenge to those rules in the *FCC v. Midwest Video Corporation (Midwest II)* case in 1979, the Court found that such rules required cable systems to operate like COMMON CARRIERS and were illegal, in contradic-

tion to the COMMUNICATIONS ACT OF 1934. Because the FCC could therefore not control or require public access, cable operators believed they had no legal responsibility to allow public access. The situation was addressed and clarified to some degree by new legislation in the CABLE COMMUNICATIONS POLICY ACT OF 1984 and the creation of PEG CHANNELS and, to a lesser extent, CUPU LEASED ACCESS CHANNELS.

Mind Extension University (ME/U) a BASIC CABLE network devoted solely to DISTANCE EDUCATION. The programming consists of high school courses, college credit and noncredit TELECOURSES, and continuing education for personal enrichment and professional development.

minimum advertised price (MAP) a price representing a threshold for the retailer advertising a MAP video title that is designated as such. It is established by the PROGRAM SUPPLIER. The retailer is free to lower that price but a store that advertises a title below the MAP is ineligible to collect CO-OP ADVERTISING reimbursement for the ad and therefore benefits by adhering to the original price. MAP is a questionable and legally ambivalent practice, inasmuch as it can be construed by the FEDERAL TRADE COMMISSION (FTC) as a tactic that fosters unfair competition by maintaining minimum sale prices. Price-fixing at the retail level is against the law and MAP may be construed as price fixing in some circumstances.

miniseries a short-form dramatic television program. The concept grew out of the need for more time to tell a story than was available in a two-hour or even a two-part four-hour MADE-FOR-TV MOVIE. Initially, such programs were called limited series. As this type of program evolved, however, the term "miniseries" came into favor. It more specifically describes a short string of sequential programs that are televised within a week's time, usually on consecutive evenings. Both limited series and miniseries are prescheduled, designed to run to their conclusion, and are not cancelled before the completion of the story. "Roots" is the most successful miniseries to date. It was first telecast over eight evenings in 1977.

The ACADEMY OF TELEVISION ARTS AND SCIENCES (ATAS) inaugurated a Limited Series category for the EMMY AWARDS in 1974. The nomenclature was changed to Mini Series (since revised to Miniseries) in 1986.

Minitel See TELETEL.

MM an abbreviation for one million, often used in ADVERTISING. It expresses a rate or price based on one million units. The designation, however, is not accurate. The Roman numeral for 1,000 is M, and MM in Roman numerals is 2,000. Common industry usage has adulterated the meaning of MM, however, to indicate one million.

The initials mm (in lower case) are also used in film production to indicate size in millimeters, such as 16mm or 35mm film and to designate the 8MM VIDEO FORMAT. (See also M.)

modulation a process that involves the alteration of the FREQUENCY phase or AMPLITUDE of an electronic signal. It involves the imposition of a pattern of variations on a straight stream of energy. A device called a modulator changes the signal and impresses a new signal or pattern on a steady carrier wave. The process of modulation always causes a slight degradation of the original signal. (See also AMPLITUDE MODULATION [AM] and FREQUENCY MODULATION [FM].)

Mom-and-Pop video store a small local video neighborhood store, often operated as a family enterprise. The independently owned establishments usually contain some 2,000 square feet of retail space and offer approximately 4,000 different titles. They are sometimes operated as a FRANCHISE or AFFILIATE of a national operation. (See also BUYING GROUPS and VIDEO SOFTWARE DEALERS ASSOCIATION [VSDA].)

monitors special television display units used in professional television operations to observe or measure program material. They usually accept only direct video signals (no RF signals from ANTENNAS) and do not have audio capabilities. Banks of small monitors are RACK-MOUNTED in control rooms and are used by directors to choose shots and by engineers to monitor the quality of the picture. They are often monochrome receivers, even though the picture may be in color, because black-and-white monitors usually provide a sharper image.

Other types of monitors, often with larger screens, are used for sales presentations in CORPORATE TELEVISION or in AUDIOVISUAL COMMUNICATIONS. Mounted on movable carts, these monitors offer superior pictures (often cleaner and more detailed than regular TV pictures) as well as better sound. The term is used in the television industry to specifically denote the utilization of a very high-quality receiver for a specific purpose in a professional setting. (See also VIDEOWALLS.)

moratorium   a marketing scheme in which the home video PROGRAM SUPPLIER offers a prerecorded video title for a limited time and then withdraws it from the market. The purpose is to create demand during that short period. The suppliers engage in the tactic by releasing a title and placing a firm cutoff date for orders, but the scheme only works if the date is kept and no other orders are allowed. Titles released under the moratorium method are returned to the market after two or three years.

Motion Picture Association of America (MPAA)   a private association that represents the motion picture industry in the United States before Congress and the public. It is composed of the major Hollywood film studios. The association is best known to the public as the creator and administrator of the MOVIE RAT-

INGS SYSTEM (R, PG, G, etc.), but it also represents the Hollywood film industry abroad with its Motion Picture Export Association of America.

The Movie Channel (TMC)   a PAY (PREMIUM) CABLE network that is devoted exclusively to first-run theatrical full-length motion pictures. It began transmitting via SATELLITE in 1979.

movie rating systems   organized procedures for classifying motion pictures according to content. The major rating system for theatrical films was inaugurated in 1968 to assist parents in choosing movies for their children. Today it is administered by the MOTION PICTURE ASSOCIATION OF AMERICA (MPAA). The RATINGS are: G (General Audience - all are admitted); PG (Parental Guidance Suggested - some material may not be suitable for children); PG-13 (Parental Guidance Suggested - no one under 13 admitted); R (Restricted - youths under 17 must be accompanied by a parent or adult guardian); and X - no one under 17 admitted.

The system is voluntary—producers may or may not choose to submit their films to the MPAA for a rating. There are no written guidelines and the films are rated by theatrical (not television, cable, or video) standards. A panel of parents rates the film on its content, not its quality. Most of the film, television, cable, and video industry has adhered to the MPAA standards. The ratings are often published in television and cable program guides and are usually placed on videocassette boxes and on advertising and POP (point of purchase) displays.

Another rating system has been developed by the Film Advisory Board (FAB), a Los Angeles-based group of producers and interested citizens. Its system has been adopted by some independent film makers and by those who believe the MPAA ratings do not go far enough in describing a film. The FAB has six major designations: C-Children through age 7; F - Family; M - Mature; VM - Very Mature; EM - Extremely Mature; and AO - adults 18 and older. In addition the FAB-printed labels on cassette boxes add descriptions such as "frontal nudity," "extreme language," "substance abuse," "violence," and "erotica."

multicam system an early method of recording television shows. Three film cameras were positioned in the studio and shot the show from different angles. They could be turned off or on at will but in many cases were simply kept running for the entire show. The resulting films were then edited, using three connected Moviola machines. The quality of the finished film was vastly superior to the KINESCOPE process.

multichannel multipoint distribution service (MMDS) a form of PAY TV. Sometimes called "multichannel television (MCTV)" and more recently "wireless cable," MMDS offers viewers a choice of programming for a fee, similar in nature to BASIC and PAY (PREMIUM) CABLE SERVICES. It has recently gained the capability of offering more than

one channel of programming and therefore differs from the LOW POWER TELEVISION (LPTV) stations and the now-defunct SUBSCRIPTION TELEVISION (STV) stations, with their single broadcast channels.

The transmission system operates from a 10-watt (or with a variance, a 100-watt) microwave transmitter, using the two-GIGAHERTZ (GHZ) band to broadcast omnidirectional signals in a radius of 15 to 25 miles. The signal can be received at most locations within the radius, allowing for a number of receiving points. Wireless cable usually transmits an unscrambled signal because regular home antennas cannot pick up the high microwave transmissions. The technology is the same as that used in INSTRUCTIONAL TELEVISION FIXED SERVICE (ITFS) and OPERATIONS FIXED SERVICE (OFS) systems.

Reception is provided by the installation of a special antenna on the rooftop of the household. A box on the back of the antenna receives the signal and down-converts it to a standard VHF frequency and sends the signal by cable to the subscriber's television set. The subscriber pays a monthly fee to receive the signal.

multiformat release the release of a home video title in more than one VIDEOTAPE FORMAT. The program or film is released simultaneously in VHS and BETA and perhaps on 8MM VIDEO FORMAT. On occasion the title is also released at the same time on videodisc.

Multilingual Communications Association (MLCA) a nonprofit trade association formed in June 1991 to inform the government, media, academia, and the public about non-English television programming. It promotes the export to other countries of non-English programs that have been produced in the United States.

multimedia the integration of more than one medium for an educational purpose. A 16mm projector, a videocassette machine, 35mm slide machines, or other devices are linked together, often by a personal computer (PC). The technique was given its formal name in the 1950s by AUDIOVISUAL COMMUNICATIONS professionals.

Sometimes called a "cross media" or "mixed media" approach to instruction, the original method involved the classroom use of books, globes, realia, records, audio tapes, charts, films, slides, and other teaching devices. The projected images in the mix were often seen on sections of a big screen, controlled by the teacher or professor, and used in a large-group environment. Today, the multimedia approach has been individualized and built around a computer. Text, graphics, still pictures, animation, sound, and video are incorporated into a videodisc in a more nonsequential, nonlinear fashion. They are at the beck and call of the individual learner at a workstation in an INTERACTIVE MULTIMEDIA system.

The term multimedia is also used in the advertising world, where it applies to the simultaneous use of more than one type of communications technology in an advertising CAM-PAIGN. (See also HYPERMEDIA and INTERACTIVE MULTIMEDIA ASSOCIATION [IMA].)

multimedia buy See CROSS-MEDIA BUY.

multipay See PAY (PREMIUM) CABLE SERVICE.

multiple dwelling units places of residence with more than one dwelling unit. The term describes apartment houses, hotels, and condominiums. Such units are often served by private cable operators rather than cable franchises. (See also SATELLITE MASTER ANTENNA TELEVISION.)

multiple system operator (MSO) a company that owns and operates more than one cable system. The firm has been awarded or has acquired FRANCHISES in a number of locations and operates all of them in a coordinated manner. They are the GROUP BROADCASTERS of cable. MSO firms can economize operations through centralized management and volume discounts on equipment and supplies. In addition, the high number of subscribers in cable systems owned by a large MSO means that programming from BASIC CABLE networks can be obtained at lower rates. The largest MSOs have ties to or own parts of production companies, cable networks, and other program suppliers and are said to be "vertically integrated" in the industry. There are no federal rules or regulations regarding the number of cable systems that can be owned by any one company. Some 350 MSOs are in operation in the United States and there has been an increasing trend toward further consolidation. The 10 largest MSOs have approximately 45 percent of the cable subscribers.

multiplexer See FILM CHAIN.

multitap a device used in cable systems to select portions of the signal from the FEEDER CABLES to serve more than one subscriber from a single location. This electronic component is usually mounted at a telephone pole location and can provide service to two, four, or eight subscribers. It taps signals from the feeder cable that are then sent to each subscriber's home by CABLE DROP lines.

Museum of Broadcasting (MB) See MUSEUM OF TELEVISION AND RADIO.

Museum of Broadcasting Communications (MBC) a Chicago-based organization dedicated to the display of broadcasting memorabilia. Unlike the MUSEUM OF TELEVISION AND RADIO in New York, MBC is more an entertainment than a serious research center. The two are separate entities, with no official connection. In addition to housing more than 3,500 television and 1,000 vintage radio broadcasts, MBC also allows visitors to participate in a mock newscast, displays a re-creation of Fibber McGee's closet, and features a small kiosk where one can see the "100 Funniest Commercials." (See also

AMERICAN MUSEUM OF THE MOVING IMAGE [AMMI], ATAS/UCLA TELEVISION ARCHIVES, LIBRARY OF CONGRESS: MOTION PICTURE, BROADCAST, RECORDED SOUND DIVISION, and NATIONAL CENTER FOR FILM AND VIDEO PRESERVATION.)

Museum of Television and Radio a nonprofit institution, located in New York, that collects, interprets, and exhibits radio and television programs. The museum houses copies of 15,000 radio programs, 25,000 television programs, and 10,000 commercials covering more than 70 years of broadcasting history. The facility contains two theaters, two large-screen listening rooms, and three exhibition galleries. In addition there are 95 television screening consoles and 25 radio listening posts, each equipped with a videocassette or audio player, where students, scholars, and the public can view or hear individually selected programs. (See also AMERICAN MUSEUM OF THE MOVING IMAGE [AMMI], ATAS/UCLA TELEVISION ARCHIVES, LIBRARY OF CONGRESS: MOTION PICTURE, BROADCAST, RECORDED SOUND DIVISION, MUSEUM OF BROADCASTING COMMUNICATIONS [MBC], and NATIONAL CENTER FOR FILM AND VIDEO PRESERVATION.)

must-carry rules FEDERAL COMMUNICATIONS COMMISSION (FCC) rules requiring cable systems to carry the signals of television broadcast stations. Systems with 12 or fewer channels must program a minimum of three local stations. Systems with more than 12 channels are required to reserve up to one-third of their capacity for broadcasters. PUBLIC TELEVISION (PTV) stations are automatically carried.

The rules are coupled with retrans- mission consent rules, which give the right to commercial broadcasters to negotiate with cable companies to receive remuneration for the carriage of their signals.

N

NAB Code   See TELEVISION CODE.

narrowband   a communications system that utilizes a narrower and lower FREQUENCY range, compared to the higher wideband services. A service using frequencies below 1 MEGAHERTZ (MHZ) is considered a narrowband service, whereas television at 6 MHZ is commonly considered a wideband service.

narrowcasting   a buzzword from the 1970s referring to the capabilities of the many new technologies to communicate with carefully targeted audiences and DEMOGRAPHICS. The opposite of BROADCASTING, the word evoked images of entertainment or educational programs designed for very specific purposes to reach a number of discrete, particular, limited-interest viewers.

It exemplified the opportunities that were possible in moving from the mass communications MEDIA such as radio and television to more personal and intimate forms of communications such as cable and video-cassettes.

While the term is still used in some circles, it lost its cachet as a futuristic expression when the new technology and its many uses became more commonplace in the late 1980s. (See also FRAGMENTATION.)

National Academy of Cable Programming (NACP)   See NATIONAL CABLE TELEVISION ASSOCIATION (NCTA).

National Academy of Television Arts and Sciences (NATAS)   New York-based national, professional organization responsible for administering the annual EMMY AWARDS in sports, news and documentaries, and daytime programming along with some technical awards. The nonprofit membership organization coordinates and assists 17 local chapters in major cities in the United States, which award their own Emmys.

Prior to 1976 NATAS was the only organization devoted to recognizing excellence in television. In that year, the West Coast group withdrew from NATAS and founded the ACADEMY OF TELEVISION ARTS AND SCIENCES (ATAS) that is now responsible for administering the Emmys for achievement in nighttime programming, along with some technical awards.

National Asian American Telecommunication Association (NAATA) a multicultural arts organization based in San Francisco, NAATA's mission is to produce, promote, and present works in film, video, and radio, by and about Asians and Asian-Americans. NAATA was formed in 1980 by Asian-Americans active in media.

National Archives and Records Administration: Motion Picture Sound/Video Branch a branch of the federal government that houses one of the world's largest audiovisual archives. The extensive collection includes DOCUMENTARIES, newsreels, and raw historical footage. Government films as well as commercial productions are included in the 120,000-film and 13,000-video collection. (See also AMERICAN MUSEUM OF THE MOVING IMAGE [AMMI], ATAS/ UCLA TELEVISION ARCHIVES, LIBRARY OF CONGRESS: MOTION PICTURES, BROADCAST, AND RECORDED SOUND DIVISION, MUSEUM OF TELEVISION AND RADIO, MUSEUM OF BROADCASTING COMMUNICATIONS [MBC], and NATIONAL CENTER FOR FILM AND VIDEO PRESERVATION.)

National Association of Black-Owned Broadcasters (NABOB) a nonprofit association, headquartered in Washington D.C., that consists of black owners of radio and television stations, individuals who hope to become involved in ownership, adver-

tisers interested in reaching the black community, and related professional associations and communications schools. Founded in 1976, NABOB represents black-owned stations before Congress, the FEDERAL COMMUNICATIONS COMMISSION (FCC), and other governmental agencies. The group holds semiannual meetings.

National Association of Broadcast Employees and Technicians (NABET) a trade union that represents technicians and engineers at television operations throughout the country. It includes camera operators, electrical workers, GRIPS, art directors, costume designers, set construction and wardrobe people, and audio and video engineers in both film and television. It is an affiliate of the American AFL–CIO and its Canadian counterpart. The union represents employees at NBC and ABC, but not employees at CBS, who are represented by the INTERNATIONAL BROTHERHOOD OF ELECTRICAL WORKERS (IBEW).

National Association of Broadcasters (NAB) the largest TRADE ASSOCIATION in the broadcast industry. Founded in 1923, it is based in Washington D.C. and represents both radio and television stations and networks before Congress and the public. The NAB is also prominent in protecting commercial broadcasting interests before the FEDERAL COMMUNICATIONS COMMISSION (FCC). For a number of years, it operated the Television Information Office (TIO) in New York City and it continues to house and support the BROADCAST

PIONEERS LIBRARY in Washington D.C. The NAB annual convention is the largest gathering of broadcasters in the world.

National Association of College Broadcasters (NACB) a trade association of college radio and television stations that concentrates on the needs of student-staffed facilities. The organization was founded in 1988 at Brown University and draws its membership from the approximately 400 college and university student stations that reach their campuses and university communities via cable.

National Association of Media Brokers (NAMB) a nonprofit organization that compiles data on the buying and selling of MEDIA properties, publishes studies, and hosts a semiannual conference. It was founded in 1979 and is headquartered in New York City.

National Association of Public Television Stations (NAPTS) See AMERICA'S PUBLIC TELEVISION STATIONS (APTV).

National Association of Radio and Telecommunications Engineers (NARTE) a nonprofit organization that seeks to encourage greater professionalism in the electronic engineering field along with education in communications in colleges and universities. The association creates certification guidelines, holds engineering seminars, and hosts an annual conference.

National Association of Recording Merchandisers (NARM) a nonprofit TRADE ASSOCIATION that provides research and member services for companies involved in the recording industry including phonograph records and audiotapes. The association was the moving force in the development of the home video industry and became the founder of the VIDEO SOFTWARE DEALERS ASSOCIATION (VSDA). The two associations became separate entities in 1991.

National Association of Regional Media Centers (NARMC) an affiliate of the ASSOCIATION FOR EDUCATIONAL COMMUNICATIONS AND TECHNOLOGY (AECT). The membership group fosters the exchange of ideas among educational media specialists at regional media centers.

National Association of State Educational Media Professionals (NASTEMP) an organization composed of AUDIOVISUAL COMMUNICATIONS professionals who work for state offices of education. It encourages the use of media in instruction at the K–12 level. NASTEMP is affiliated with the American Association of School Libraries and the ASSOCIATION FOR EDUCATIONAL COMMUNICATIONS AND TECHNOLOGY (AECT).

National Association of Telecommunications Officers and Advisors a nonprofit association that consists of officers and executive directors of local government franchise authorities in the cable industry. It is an affiliate of the National League of Cities.

National Association of Television and Electronic Service Associates See NATIONAL ELECTRONICS SALES AND SERVICE DEALERS ASSOCIATION (NESSDA).

National Association of Television Program Executives See NATPE INTERNATIONAL.

National Association of Video Distributors (NAVD) a nonprofit TRADE ASSOCIATION that is primarily composed of 62 WHOLESALERS or distributors in the home video programming business. Manufacturers, PROGRAM SUPPLIERS, and others in the field are associate members.

National Audio-Visual Center (NAVC) a federal agency that is the repository and distributor of all government-produced films and video programming. Based in Washington, D.C., it provides copies of most of the titles produced by all federal agencies or departments to the public and to schools, colleges, and libraries. The agency is a part of the General Services Administration (GSA) and also conducts an annual survey for Congress on the amount of film and television production completed by federal government agencies and the number of television and film facilities operated by those agencies.

National Black Media Coalition (NBMC) a nonprofit organization that monitors the MEDIA and promotes civil rights and the representation of minorities in all media

including broadcast and cable television. It was established in Washington D.C. in 1973.

National Black Programming Consortium (NBPC) an organization that assists PUBLIC TELEVISION (PTV) stations in expanding their programming. It collects and archives black-oriented television programming, co-produces black programming, and helps financially in the acquisition and distribution of that programming.

National Broadcasting Company (NBC) the oldest of the three major commercial full-service television networks. It is headquartered in New York City. The company was incorporated in 1926 as the National Broadcasting Company by a consortium consisting of Westinghouse, General Electric, and the RADIO CORPORATION OF AMERICA (RCA), who were then partners in RCA. It was the first company organized specifically to operate a broadcast network. In 1930 RCA took over control of the network and by 1932 it had become the sole owner.

NBC became the most popular radio network, a position it held until the early 1940s. NBC also pioneered in television, beginning experimental transmissions in 1931 and inaugurating limited but regular television service on April 30, 1939 with a broadcast from the New York World's Fair featuring President Franklin D. Roosevelt. World War II stopped further television development, but after the war NBC was the leader in developing the new medium. In those early days, NBC captured nearly all of the nation's few viewers with comedian Milton Berle who became known as Mr. Television. Later, the network introduced the TALK SHOW format and pioneered the SPECTACULAR or SPECIAL programming concept in PRIME TIME. The network also initiated the participating sponsorship of programs and introduced the television audience to the MINISERIES.

Perhaps NBC's most significant contribution to the communications industry was its pioneering role in fostering and promoting color television programming. The first NBC color telecast was the Pasadena Tournament of Roses on New Year's Day in 1954. The network gradually increased its color coverage to include all of its programs and the two other major networks eventually followed.

In 1986 the company's parent (RCA) was acquired by the General Electric (GE) Company (one of the original founders of the network), and NBC is now a subsidiary of GE. The corporation operates six O & O television stations and also has interests in cable programming operations including the ARTS AND ENTERTAINMENT (A&E) NETWORK and BRAVO, SPORTSCHANNEL AMERICA, and its newest venture, the CONSUMER NEWS AND BUSINESS CHANNEL (CNBC). The company that originated network radio, however, sold its last radio station in 1989.

National Broadcasting Society
See ALPHA EPSILON RHO (AERHO).

National Cable Television Association (NCTA) the largest trade association in the cable industry. Founded in 1952, the membership of the nonprofit organization consists of companies involved or associated with the industry, including individual system operators, MULTIPLE SYSTEM OPERATORS (MSO), equipment manufacturers, cable brokers and financial companies. The NCTA represents cable interests in dealings with Congress and before the FEDERAL COMMUNICATIONS COMMISSION (FCC) and within the communications field, proposing and supporting legislation and regulatory action beneficial to the cable industry.

The Washington D.C.-based association engages in research on behalf of its members and maintains an active promotion and publication program concerning cable. Since 1985 the NCTA has operated the National Academy of Cable Programming (NACP), the sponsor of the ACE AWARDS, which are presented for excellence in programming. The NCTA annual convention and trade show is the largest in the industry.

National Cable Television Cooperative Inc. (NCTC) a nationwide member-owned purchasing cooperative. The goals of the NCTC are to achieve favorable pricing and terms in the purchase of hardware and programming by functioning as a purchasing agent for the collective membership.

National Cable Television Institute (NCTI) an independent training organization that specializes in self-study programs in the technical area of the cable television industry. NCTI trains cable television technicians and engineers at individual cable systems, MULTIPLE SYSTEM OPERATORS (MSO), installation contracting companies, and cable industry vending firms.

National Captioning Institute (NCI) a nonprofit corporation that has developed a national CLOSED-CAPTIONED television service for the entertainment industry. Funded by grants from the Department of Health, Education, and Welfare and the television networks, the institute is now partially self-sustaining. The primary purpose of closed captioning is to provide deaf and hearing-impaired television viewers with captioned dialogue. NCI is involved in producing captions and marketing decoders.

National Center for Film and Video Preservation an organization designed to discover and preserve film and television programs for inclusion in the AMERICAN FILM INSTITUTE (AFI) collection at the Library of Congress. Funded by grants to the AFI from foundations and the National Endowment for the Arts, the center has created the computerized National Moving Image Data Base (NAMID), which lists the film and video holdings in archives throughout the United States. (See also AMERICAN MUSEUM OF THE MOVING IMAGE [AMMI], ATAS/UCLA TELEVISION ARCHIVES, LIBRARY OF CONGRESS:

MOTION PICTURE, BROADCAST, RECORDED SOUND DIVISION, MUSEUM OF BROADCASTING COMMUNICATIONS [MBC], and MUSEUM OF TELEVISION AND RADIO.)

National Council of Churches: Broadcasting Commission an ecumenical agency that is a cooperative organization of 19 Protestant and Orthodox denominations and agencies in broadcasting, film, cable, and the print media. It is officially known as the "Communication Commission of the National Council of Churches of Christ in the USA." It offers criticism and support on media issues to the government and the industry at large and acts as a liaison between the Council of Churches and the networks, the media, and television stations and cable systems.

National Electronic Distributors Association (NEDA) a nonprofit association that consists of wholesale distributors of electronic components. The organization sponsors research and studies, develops reports, and provides marketing information for its members.

National Electronics Sales and Service Dealer Association (NESSDA) a nonprofit organization composed of electronic service organizations at the local and state levels as well as small individual dealers. The association offers apprenticeship certification and training programs through the INTERNATIONAL SOCIETY OF CERTIFIED ELECTRONIC TECHNICIANS (ISCET), publishes magazines and newsletters, and cohosts (with ISCET) an annual convention called the National Professional Electronics Convention and Trade Show (NPEC).

National Federation of Local Cable Programmers (NFLCP) a nonprofit organization that is dedicated to providing information and services to personnel working in LOCAL ORIGINATION at cable systems throughout the United States as well as to citizens who seek to use cable and participate in the medium. The group's membership includes people in government, libraries, and other nonprofit groups that develop local programming for their cable systems, usually using PEG CHANNELS.

National Film Board of Canada (NFB) a cultural agency of the Canadian federal government. The NFB was founded by an Act of Parliament in May 1939 to initiate and promote the production and distribution of films in the national interest, with the primary object of interpreting Canada to Canadians and to other nations. It maintains 12 audiovisual centers in Canadian cities. NFB films are also shown regularly on Canadian television and cable operations and are sold in more than 80 countries.

National Film Registry an agency of the U.S. government. The registry was created by Congress in 1988 to honor, protect, and preserve theatrical films that are considered "culturally, historically, or aesthetically significant." The films are selected from hundreds nominated by the public and preliminarily recommended by the National Film Preservation Board, made up of representatives of the ACADEMY OF MOTION PICTURE ARTS AND SCIENCES (AMPAS), the AMERICAN FILM INSTITUTE (AFI), the MOTION PICTURE ASSOCIATION OF AMERICA (MPAA), the DIRECTORS GUILD OF AMERICA (DGA), and other organizations. The registry is located in Washington, D.C. and operates under the supervision of the Library of Congress, whose head makes the final selections.

National Information Center for Educational Media (NICEM) a profit-making organization that catalogs and electronically stores information about all types of software materials used in AUDIOVISUAL COMMUNICATIONS. The data base contains descriptions of films, filmstrips, and audio and videotapes and is available in book form or as an online service.

National Moving Image Data Base (NAMID) See NATIONAL CENTER FOR FILM AND VIDEO PRESERVATION.

National Public Broadcasting Archives a part of the University of Maryland library service that houses the written records and oral history recordings of the noncommercial broadcasting industry. Inaugurated under the auspices of the Academy for Educational Development in Washington, D.C. in 1990, the archives are located at the College Park campus.

National Religious Broadcasters (NRB) a nonprofit association that serves as a basic source of information about Christian broadcasting. Formed in 1944, it is composed of religious radio and television stations and the producers of religious programs.

National Technological University (NTU) a private graduate school that has delivered credit and noncredit TELECOURSES and TELECONFERENCES in computer science and engineering to 118 member institutions since 1984. It administers the nation's first satellite-delivered master's program via DISTANCE EDUCATION.

National Telecommunications and Information Administration (NTIA) an agency within the U.S. Department of Commerce that is responsible for advising the executive branch of government on telecommunications issues. It represents the views of the president on telecommunications policies before the FEDERAL COMMUNICATIONS COMMISSION (FCC) and other governmental agencies. The term "telecommunications" in the NTIA lexicon includes data communications, telephone, radio and television broadcasting, air and sea radio, and other related technologies.

National Telemedia Council a nonprofit organization that works to improve the quality of radio and television programming by educating viewers about the media. Founded in 1953 as the American Council for Better Broadcasts, the membership organization consists of local, state, and national citizens' groups and individuals. It is based in Madison, Wisconsin.

National Television Systems Committee See NTSC.

National Translator Association (NTA) a nonprofit association that consists of owners and operators of FM or TV TRANSLATOR STATIONS, mostly in the Rocky Mountain states. Its members share information about technical and regulatory matters.

National University Consortium for Telecommunications in Teaching See INTERNATIONAL UNIVERSITY CONSORTIUM (IUC).

Native American Public Broadcasting Consortium a nonprofit membership organization that is composed of PUBLIC TELEVISION (PTV) stations, community organizations, schools, and tribal units. Its purpose is to encourage the production and distribution of programming by American Indians.

NATPE International a trade association whose members are local television PROGRAM MANAGERS, programmers from GROUP BROADCASTERS, professionals from STATION REPRESENTATIVE companies and ADVERTISING AGENCIES, individuals from research organizations, and PRODUCERS and SYNDICATORS. The organization works to improve television programming by maintaining a venue that encourages the exchange of ideas and information. It is best known for its annual convention, which has evolved into the major marketplace for the selling and buying of syndicated programs. (See also NATPE NET.)

NATPE Net   an electronic mail system that links commercial television stations and SYNDICATORS to a computer data base at NATPE INTERNATIONAL headquarters in Los Angeles and allows for the instant exchange of information about programs among the participants. Information from other sources is also available.

NBC   See NATIONAL BROADCASTING COMPANY (NBC).

needle-down fee   a charge made by some stock sound-effects and record libraries for each use of their music or recorded effects such as train whistles and mooing cows. The owner of the copyright, which is usually the record library, receives a fee every time the work is used. It is sometimes called a "needle-drop" fee. Although most of the sounds and music are on tape today, the old record term (needle) is still used.

negative match back   a phrase used in television production editing, when the original footage was shot on film. The edge numbers on the film are transferred to the video time code on the videotape. Rough editing is completed using the videotape and the final edit is completed on film. The matching numbers make the process easy.

network   a broad term that identifies a group of television stations that are linked by MICROWAVE RELAY, telephone lines, COAXIAL CABLE, or SATELLITE in order to receive and transmit programs simultaneously. It also refers to a national organization that supplies programs to a number of cable systems by satellite. In VARIETESE, a broadcast network is often called a "web." The operations are based on the simple economy of offering the same programs to different audiences in different locales. Theoretically, any two stations that are even temporarily interconnected form a broadcast network. The FEDERAL COMMUNICATIONS COMMISSION (FCC) defines a network as "a national organization distributing programs for a substantial part of each broadcast day to television stations in all parts of the United States." Although there are REGIONAL TELEVISION NETWORKS in operation throughout the United States, the dominant networks in commercial television programming have been the AMERICAN BROADCASTING COMPANY (ABC), CBS INC., and the NATIONAL BROADCASTING COMPANY (NBC).

The desirability of reaching that large audience for advertising purposes was reinforced with the establishment of the first new national television network by FOX INC. in 1986. Created largely from independent stations, the new network is called FBC.

All cable systems in the United States are served by approximately 50 BASIC CABLE networks and some 10 PAY (PREMIUM) CABLE networks.

These organizations are not generally regulated by the FCC and do not have the same relationship with their distribution channels as the broadcast networks. Local cable systems usually purchase the program services from the cable networks at a per-subscriber fee.

The PUBLIC BROADCASTING SERVICE (PBS) is not a network *per se*. It is instead a national organization of PUBLIC TELEVISION (PTV) stations that have banded together to create and control a central agency to provide programming for transmission on the individual stations. The stations, also interconnected via satellite, pay PBS for the program service. (See also OPTION TIME, PREEMPTION, and UNWIRED NETWORK.)

network buy a practice in which COMMERCIAL TIME is purchased directly from the television or cable network on which the COMMERCIALS are to be run on all stations or systems affiliated with that network. The commercials usually run concurrently on all of the systems or stations but they can appear on different days or times, according to the needs of the advertiser. A network buy is different from buying time on a MARKET-BY-MARKET basis or on an UNWIRED NETWORK.

Network Television Association an organization that promotes the value of advertising on network television to ADVERTISING AGENCIES and their CLIENTS. The association does not sell time, promote specific DAYPARTS, or negotiate prices, but it does address such issues as CLUTTER and promotes the overall image of network television and its value to national advertisers.

new advertising a theory holding that consumers must be persuaded to buy goods through an interrelated series of advertising messages. It became popular on MADISON AVENUE in the late 1980s but it is hardly "new." In addition to traditional radio and TV commercials, newspaper and magazine print ads, and outdoor billboards, this approach uses public relations campaigns, package design, direct mail, telemarketing, and all other methods of communication.

new build the construction of a new cable system by the FRANCHISE or the extension of an already existing system into new territory in the community. Before undertaking the construction, a MAKEREADY is accomplished. A new build differs from a REBUILD, which is concerned with the physical and electronic improvement of an existing system.

new edge an all-encompassing term describing the new technology and movement that promises to revolutionize communications and entertainment in the twenty-first century. The technology includes machines that allow one to peer into another's mind, microchips inserted into the brain for instant fluency in foreign languages, CD-ROM, virtual reality, computers, and any revolutionary gadgets. The term is based in part on the New Age movement, reflecting that culture's interest in human potential and spirituality.

New World Information and Communication Order (NWICO) an ambiguous phrase describing a political issue concerning global communications. The term was used by many developing and Third World member countries of UNESCO, many of whom were critical of the widespread dissemination of western ideals and culture. They hoped to limit what they believed was the biased and overpowering influence of the developed industrialized countries on the free flow of ideas between nations. In their view, this "one-way flow" monopolizes the media and controls events, particularly in the era of satellite communications. Existing global news flow is seen as unbalanced and imperialistic. Most developed nations believe, however, that the concept would legitimatize governmental control of the press by requiring licenses for reporters. NWICO continues to be an objective of many Third World countries, but the United Nations removed NWICO as an agenda item in 1990, effectively killing it as an idea in that forum.

News Corporation Ltd., The a global communications giant headed by Rupert Murdoch. It is the second largest communications empire in the world, after TIME WARNER. The company owns some 107 newspapers on four continents, an airline, a medical publisher, 27 magazines, and the large book publisher HarperCollins in the United States and the United Kingdom.

In Australia, the company owns 100 newspapers and nine maga-zines. In the United Kingdom, the company owns five newspapers and eight magazines. The corporation owns nine U.S.-based magazines and in 1988 purchased Triangle Publications (owner of *TV Guide*, the top circulation magazine in the United States). The firm is also a partner in the BRITISH SKY BROADCASTING company, a DIRECT BROADCAST SATELLITE (DBS) service in the United Kingdom, and was a partner in another DBS service, Sky Cable in the United States. The company is incorporated in Australia but operates primarily in the United Kingdom and the United States.

The News Corporation Ltd. is best known in television in the United States as the parent corporation of FOX INC. It also owns the motion picture studio, 50 percent of the home video company CBS/Fox Video, two television production and SYNDICATION firms (Twentieth TV and Twentieth Century Fox), seven television stations, and the nation's fourth television network, FBC (Fox Broadcasting Company).

news-talk programming a radio program FORMAT that found its way quite naturally to television. It was first seen by a television audience in Chicago in 1949, when Dave Garroway hosted a news-interview-comedy show titled "Garroway at Large." NBC perfected the format with the early morning, two-hour "Today" show (with Garroway as host), which began broadcasting live from New York in 1952. The comedy skits were dropped but the news and celebrity interviews and the feature stories, along with short hard newscasts (including weather updates), were retained. The popular format survives today with many imitators, both locally and nationally. While many of the on-air discussions are about a particular news topic, others are soft features or interviews with celebrities or personalities who use the opportunity to PLUG a new book or movie, and there are occasional performances by visiting entertainers. In this way the shows using this format closely resemble TALK-SHOW PROGRAMMING and it is sometimes difficult to distinguish between the two.

Currently the purest version of the news-talk format is the "MacNeil/Lehrer Newshour" on the PUBLIC BROADCASTING SERVICE (PBS). In this program, the newscasters report on the day's events and then turn to studio guests for in-depth interviews about a specific topic. The emphasis is on breaking news, and feature stories are used only on an occasional basis. Most news-talk programs, however, blend light entertainment and information with an emphasis on talk and discussion rather than on hard news reporting.

Nielsen Homevideo Index (NHI) research reports produced by the A. C. NIELSEN COMPANY. The name is somewhat misleading. Most of the reports in this service concentrate on RATINGS and SHARES developed for cable networks and SUPERSTATIONS, not videocassette recording devices. The quarterly Nielsen Cable Activity Report (NCAR) provides national ratings for cable networks. The Cable Audience Profile (CAP) provides viewing reports on cable systems and the Cable On-Line Data Exchange (CODE) service offers information about all of the cable systems in the United States.

Nielsen Station Index (NSI) an audience rating service of the A.C. NIELSEN COMPANY that concentrates on the local television station's audience in DESIGNATED MARKET AREAS (DMA) on an individual market basis. It measures the size of the audience in more than 200 individual markets in the United States by combining information from 100,000 household diaries and 4,000 PEOPLE METERS. The service is complementary to the NIELSEN TELEVISION INDEX (NTI). The Nielsen Broadcast Index (NBI) provides a similar service for local stations in Canada. (See also DIARY SYSTEM.)

Nielsen Syndicated Service (NSS) audience estimates made by the A. C. NIELSEN COMPANY for certain syndicated programs. The NSS tracks some 450 syndicated programs sold on the BARTER basis, including both regularly scheduled series as well as SPECIALS. The findings are tabulated in weekly POCKETPIECE reports.

Nielsen Television Index (NTI) an audience rating service of the A. C. NIELSEN COMPANY. It concentrates on NETWORK television programming, measuring the size of the national audience by the use of PEOPLE METERS. The company also provides a similar service in Canada for the national networks in that country.

Nipkow scanning disc a MECHANI-CAL TELEVISION scanning disc designed to transmit images by wire in 1883. It was developed by the German inventor Paul Nipkow. The device that he called an "electrical telescope" is considered the first to reproduce images of moving objects. A primitive camera was focused on an object and a disc with holes in it was placed between the object and a lamp and rotated at a rapid rate. This varying light intensity was received through a similar disc rotating in a synchronous manner. A primitive image of the object (only inches wide) was created in a viewer's mind by the phenomenon of PERSISTENCE OF VISION. The mechanical system dominated research in television until the 1930s when it was gradually replaced with experiments in electronic television transmission and reception.

nonbroadcast television See COR-PORATE TELEVISION.

noncommercial television See EDUCATIONAL TELEVISION (ETV) and PUBLIC TELEVISION (PTV).

nonduplication rules FEDERAL COMMUNICATIONS COMMISSION (FCC) rules effective in 1990 that give some protection to broadcasters by reinstating variations of the SYNDICA-TION EXCLUSIVITY RULES and modifying its nonduplication rules concerning network programming. Under the rules local broadcasters are entitled to demand that a cable system delete network programs from distant sources that duplicate the programs on their stations. A station must, of course, have the right to the exclusive showing of the program in its own local geographic area and thus the nonduplication rules parallel the syndicated exclusivity rules. A local station can therefore block the importation of a network program from a distant station at any time, if both plan to carry the program.

nonprint materials See AUDIOVI-SUAL COMMUNICATIONS.

nontheatrical films/programs/videos any film, program, or video that is not dramatic and intended for a mass audience. The term originated in the 1930s to describe films that were not intended for release to motion picture theaters. Most of them were produced and distributed on 16mm film and were designed for the educational market.

Today, DOCUMENTARIES and corporate, training, experimental and artistic films as well as children's programming, HOW-TOs, educational films or videos, and SPECIAL INTEREST (SI) PROGRAMMING are nontheatrical. They are often created by INDEPEN-DENT PRODUCTION COMPANIES for a narrow TARGET AUDIENCE. (See also COUNCIL ON INTERNATIONAL NONTHE-ATRICAL EVENTS [CINE].)

North American Broadcast Teletext Standards (NABTS) a one-way TELETEXT information system that seeks to become the technical teletext standard in the United States. Developed by the AMERICAN

TELEPHONE AND TELEGRAPH COMPANY (AT&T) and initially called the "presentational level protocol syntex (PLPS)," the system is compatible with both the Canadian TELIDON system and the French ANTIOPE system. Because the FEDERAL COMMUNICATIONS COMMISSION (FCC) has declined to set a U.S. standard, the marketplace will decide if NABTS, WORLD STANDARD TELETEXT (WST), or another system will become the *de facto* teletext standard in the United States.

novelas See *TELENOVELAS.*

NTSC a television engineering standard adopted by the United States in 1941, which established the technical aspects of television transmission and reception in the nation. The acronym stands for National Television Systems Committee. The FEDERAL COMMUNICATIONS COMMIS-

SION (FCC) established 525 SCANNING LINES per FRAME, 30 frames per second, and the 3:4 ASPECT RATIO for a television screen as a national standard for television. The NTSC color standard was adopted by the FCC in 1953.

Only one-quarter of non-American television systems use the NTSC standard, which is often derisively called the "never twice the same color" system. In the global television world, it is known as "System M." The standard is incompatible with the French SECAM and European PAL systems.

numbers industry jargon for the audience RATINGS and SHARES developed and published periodically by the major audience research services. The reports (or BOOKS) that are published contain the numerical calculations that are collectively known as "the numbers."

O

O & O initials for a television station that is "owned and operated" by one of the NETWORKS. They are also sometimes applied to the stations owned by GROUP BROADCASTERS.

OEM See CONSUMER ELECTRONICS.

off-air video recording the practice and the right of the public to

record television programs OFF THE AIR from broadcast stations for later viewing in the home. The issue was settled in the BETAMAX CASE in 1984. While that case did not address the issue of whether one can record a program from a cable network, presumably the same reasoning would apply. The Betamax case also did not address whether a recorded program could be viewed outside the home, although showing such a program to people in a theater and charging a fee for the privilege would appear to violate the COPYRIGHT ACT OF 1976. Whether such viewing, in a nonprofit educational setting where no fee is charged, is a violation of the law is not clear. (See also PUBLIC PERFORMANCE OF COPYRIGHTED VIDEO MATERIAL.)

off camera See OFF MIKE/OFF CAMERA.

off mike/off camera terms used to describe any action or sound in a television production that occurs out of the range of a microphone or camera. Voices or sounds made off camera give the impression of action happening there and references to movement or business out of camera range imply that such activity is taking or has taken place.

off-network See SYNDICATION.

off-network programs syndicated television shows that originally aired on one of the major networks. They are offered to individual stations, groups of stations, cable systems, and occasionally to MULTICHANNEL MULTIPOINT DISTRIBUTION SERVICE (MMDS) and LOW POWER TELEVISION (LPTV) stations after their network runs are completed. In any given season, four to seven network series make the transition to SYNDICATION. Because they are often stripped ACROSS THE BOARD on a daily basis by the transmitting entities, it is usually necessary for the producers to have at least 90 episodes of a series

(representing some four years of weekly network scheduling) to offer to buyers.

off the air a phrase that applies only to broadcasting through the airwaves and is used in both the radio and television industries. Television viewers may "pick up signals off the air" instead of by COAXIAL CABLE through a cable system. Teachers, researchers, and others may engage in OFF-AIR VIDEO RECORDING from television stations for later viewing. The ultimate technical nightmare for a chief engineer is when the station inadvertently goes "off the air."

offline editing a preliminary editing process often used in making an initial or rough cut of the program. This type of VIDEOTAPE EDITING is accomplished with DESKTOP VIDEO or PROSUMER equipment that is less expensive and less sophisticated than professional equipment. The process allows editors to make preliminary artistic decisions and to create a video WORKPRINT and editing script or edit decision list (EDL) before proceeding to the more expensive process of ONLINE editing.

Omega Xi Alpha See SOCIETY FOR COLLEGIATE JOURNALISTS.

omnidirectional microphone an inexpensive mike that is sometimes used in both film and television production. It is often called a "pressure mike." The microphone usually produces the most realistic sound because its pickup pattern is similar to that of the human ear. It can pick up sound from any and all directions and several such mikes are often used to cover a large group of people. Their wide range and nonselective sensitivity, however, also pick up unwanted ambient sound.

one-time-only (OTO) programs See SPECIALS.

online editing a process of VIDEO-TAPE EDITING that represents the final stage in the preparation of a prerecorded television program. Using the roughcut or video WORKPRINT created by OFFLINE EDITING and an edit decision list (EDL) or script as a guide, a final edit is made using professional equipment. The final online edit is conformed to (matched exactly to) the off-line tape. The new master tape is created using high-quality VIDEOTAPE FORMATS, along with sophisticated computerized controllers and SWITCHERS that use SOCI-ETY OF MOTION PICTURE AND TELEVISION ENGINEERS (SMPTE) time code videotape editing methods. The result is a finished, polished videotape for airing, distribution, or DUB-BING. (See also AB ROLL EDITING.)

open end in advertising, a phrase indicating the ending to a network or nationally syndicated program that is left blank for the insertion of a COMMERCIAL from a local firm. It gives a local company an opportunity to add its name, logo, and address to the national SPOT and to thereby participate in the sponsorship of a national program.

In broadcasting, the term is used infrequently to describe a show that has no scheduled completion time. An example is a discussion program, which can continue until the topic, the panel, and the audience are exhausted. Such shows have been common in the after-midnight hours on radio but the FORMAT has seldom been used in television.

Open University a pioneering DIS-TANCE EDUCATION project in the United Kingdom. It was designed to afford the opportunity for a college education at home by radio, TV, and correspondence study. The institution admitted its first students in 1971.

Although the school uses radio and television to transmit TELE-COURSES, print is the principal medium of instruction. Students spend some 85 percent of their study time reading and writing. The television and radio production is done under the supervision of the BRITISH BROAD-CASTING CORPORATION (BBC) and transmitted on BBC-2 and Radio 3. Most of the television programs are 24 minutes in length. Students talk with teachers on the telephone and occasionally interact face-to-face with professors at study centers throughout the country.

operations fixed service (OFS) a television broadcast transmission and reception service established in the early 1960s by the FEDERAL COMMU-NICATIONS COMMISSION (FCC) for commercial business purposes. It operates on the same two-GIGAHERTZ

(GHZ) microwave frequency band as the MULTICHANNEL MULTIPOINT DISTRIBUTION SERVICE (MMDS) and INSTRUCTIONAL TELEVISION FIXED SERVICE (ITFS) stations.

The FCC allocated the specific OFS channels to allow private companies to communicate with one another (or their branches) in an economical manner or to exchange images and data in a timely fashion. OFS stations operate exactly like MMDS and ITFS operations. The only difference is in their intended use.

option time a now illegal agreement between the television NETWORKS and their AFFILIATED STATIONS that reserved a portion of the day when one of them could exercise control over the program schedule. The periods were known as "network option time" and "station option time." In the early 1980s, the FEDERAL COMMUNICATIONS COMMISSION (FCC) issued rules prohibiting the practice. The reason for the prohibition was to fix legal responsibility for the control of broadcasting at the local level. (See also PREEMPTION.)

Oracle an acronym for Optical Reception and Announcements by Coded Line Electronics. It is a British, one-way TELETEXT system operated by the INDEPENDENT BROADCASTING AUTHORITY (IBA). Similar in nature to CEEFAX, the system transmits pages of information to television sets, which can be accessed by using a decoder and a keyboard. Oracle offers twice the number of pages as

Ceefax, but the system has had a difficult time finding public acceptance.

organizational television   See CORPORATE TELEVISION.

Oscar awards statuettes presented annually by the ACADEMY OF MOTION PICTURE ARTS AND SCIENCES (AMPAS) for outstanding achievement in the motion picture industry in the United States. The awards are given in various categories ranging from Best Picture to Best Director to Best Actor. They also reward technical and other behind-the-scenes talent. The awards ceremonies in Hollywood are televised internationally.

oscilloscope an electronic testing apparatus that graphically depicts an electronic signal as a function of time. The display appears on a CATHODE RAY TUBE that is overlaid with a grid, which acts as a measurement base. The tube is contained in a boxlike unit that can be connected to any piece of electronic gear emitting a signal. The incoming signal can be periodic or nonperiodic but it is expressed electronically on the grid of the oscilloscope tube as a single moving wave in a particular pattern that can then be measured and interpreted. It is often called a "scope." It is used to electronically test various pieces of equipment.

outside-the-home viewing an audience research phrase referring to the common situation in which people watch television in other-than-home environments. Advertisers and ADVERTISING AGENCIES have always known that there is a significant number of viewers in college dormitories, bars, hotels, and the workplace. These viewers have never been officially measured and they are not currently cited by account executives in selling COMMERCIAL TIME. (See also HOUSEHOLDS USING TELEVISION.)

outtakes   bits of film or videotape that are not used in the final edited program. They may contain flubs in lines or action by the actors, composition, or bad CAMERA ANGLES or movement. In Hollywood parlance, they are the scenes that are "left on the cutting room floor."

over-the-shoulder shot   a REVERSE ANGLE SHOT in which one person is seen over the shoulder of another. It is often used in dramatic programs and on interview shows. The shot concentrates the attention of the viewers on the person whose face dominates the frame and who is usually talking, while continuing to remind them that another person is involved. It is also sometimes used as a REACTION SHOT. (See also FRAMING.)

overbuild   the construction of more than one cable system in a FRANCHISE area. Because cable systems are usually granted and operate a franchise by a local governmental authority on a nonexclusive basis, it is possible that another franchise can be granted to another company at any time. The practice, however, is not common.

A second franchise is usually granted because the first operator is charging high rates to subscribers, has not made cable available to everyone in the community, or has a poor customer-service record.

overdubbing   a process in which the audio of a television show is often improved or enhanced. The recording on one sound track is dubbed over another in a SWEETENING action. Two, three, or more tracks can be overdubbed and combined in a variety of ways to create a unique and superior final sound.

overnight ratings   a service of the A. C. NIELSEN COMPANY that delivers—the next day—a rough estimate of the size of a PRIME-TIME audience for a particular program to ADVERTISING AGENCIES, producers, or NETWORKS. The day immediately following the telecast, the customer receives RATINGS from a few major cities based on data gathered from homes with PEOPLE METERS or by means of a TELEPHONE COINCIDENTAL SURVEY.

Within the industry the service is simply referred to as "the overnights."

P

Pacific Mountain Network (PMN) a PUBLIC TELEVISION (PTV) organization composed of 45 PTV stations in the Rocky Mountains and West Coast of the United States. The nonprofit network includes stations in 13 states and has its headquarters in Denver, where it operates a satellite UPLINK facility. (See also GROUP BUY.)

Pacific Rim Coproduction Association (PRCA) an international television production group that encourages the production of television programs about countries in and around the Pacific Ocean. The consortium consists of U.S. PUBLIC TELEVISION (PTV) stations, the CANADIAN BROADCASTING CORPORATION (CBC) and TV ONTARIO (TVO), the Australian Broadcasting Corporation (ABC), the Special Broadcasting System in Australia, and Television New Zealand.

Pacific Telecommunications Council (PTC) a membership organization consisting of professionals involved in telecommunications, particularly satellite utilization, in the Pacific, Asia, and North and South America. The council provides a venue for meetings among governments and commercial manufacturers and suppliers through conferences, workshops, and seminars.

paintbox See DIGITAL VIDEO EFFECTS.

PAL a television transmission and reception engineering standard used in most of Europe, the United Kingdom, and Australia. It is the most widely used standard in the world.

The acronym is derived from phase alternation line, although some have labeled it the "picture at last" system. It uses 625 SCANNING LINES and 25 FRAMES per second and is incompatible with the NTSC and SECAM systems. Some have called for a single worldwide television system and promote PAL as the "peace at last" system. (See also ADVANCED TELEVISION [ATV].)

pan the rotation of a stationary camera on its horizontal axis. The operator moves the camera slowly from left to right or from right to left, following the action or showing the panoramic view of the scene. A pan is ideally made in one smooth, continuous motion, without interruption. A swish pan is a rapid sideways movement, designed to portray action and speed. (See also FRAMING and TILT.)

pan-and-scan technique See LETTERBOXING.

pancake makeup a type of makeup that is applied as a base to the talent's face. In modest television productions, it frequently serves as the only makeup. A dry form of makeup that is applied with a damp cloth or sponge, pancake comes in a jar or a round flat container in varying shades ranging from light to dark skin tones.

parabolic microphone   a mike that is mounted facing the center of a concave dish, which acts as a reflector of the incoming sound. The bigger the reflector, the better the sound. The bowl-shaped device is often used to pick up distant sounds in a REMOTE production.

parallax   a French word of Greek origin referring to the apparent displacement of an object as seen from two different viewpoints. In film and television production, it is the difference between what is seen by the camera viewfinder and the camera lens. The slight angle of divergence between the two can create FRAMING problems in CLOSEUPS as well as KEY-STONE difficulties.

Paramount Communications Inc. a major communications company headquartered in New York. Known as Gulf & Western until the 1980s, the company's holdings include Paramount Pictures, the Simon and Schuster publishing house, Madison Square Garden, the New York professional hockey and basketball teams (New York Rangers and New York Knickerbockers), television stations, and motion picture theaters. The company was acquired in 1994 by VIACOM ENTERPRISES after a lengthy Wall Street battle.

participating sponsor   an advertiser that joins with other companies in buying COMMERCIAL TIME on a pro-gram or series of programs on a television or cable operation. Sometimes known as cosponsors, participating sponsors seek identification with a program of quality or a prestigious televised event. Some programs may have as many as ten participating sponsors. (See also SPONSOR.)

patch panel   power units that provide input and output connections for audio, video, and lighting equipment. They are affixed to a wall or are RACK-MOUNTED and contain rows of sockets resembling an old-fashioned telephone switchboard. Sometimes called a "patch bay," the audio and video panel allows various pieces of equipment to be connected, using standardized patch cables, and eliminates a confusing tangle of cable, cords, and wires in a television operation. Lighting patch panels are used to connect different lighting instruments to a DIMMER panel.

pay (premium) cable service   cable channels that are offered to subscribers in addition to those in its BASIC CABLE SERVICE. In order to receive them, however, the customer must first subscribe to the basic service. HOME BOX OFFICE (HBO) fully developed the concept in 1972 by offering the subscriber a special extra channel that was free of advertising.

Cable systems initially charged the subscriber separately for each new pay-cable channel ordered, but most have developed a system of packaging by TIERING. Subscribers have their choice of groups of commercial-free channels for a monthly fee.

pay per view (PPV)   a form of PAY (PREMIUM) CABLE SERVICE or DIRECT BROADCAST SATELLITE (DBS) in which the viewer is charged a single usage fee to see a specific program, motion picture, or special event. The practice is possible through the installation of an ADDRESSABLE CONVERTER in a cable customer's home that allows the cable operator to send programs via FIBER OPTICS or COAXIAL CABLE to specific viewers who have paid (or will pay) a fee. Pay per view service is also possible using satellite technology in combination with the telephone and TELEVISION RECEIVE ONLY (TVRO) dishes.

pay TV   the concept of charging the public directly for receiving programming. The idea of requiring direct payments to receive programs was initiated when television was developed after World War II. Labeled "toll TV," the precursors of SUBSCRIPTION TELEVISION (STV), PAY (PREMIUM) CABLE SERVICES, and BASIC CABLE SERVICES were launched in the early 1950s. Communication entrepreneurs experimented with both over-the-air and cable types of pay TV.

Broadcast pay-TV experiments were reluctantly authorized by the FEDERAL COMMUNICATIONS COMMISSION (FCC), because the Commission did not want to jeopardize the growth of the fledgling television broadcast medium. There was indeed considerable pressure from commercial broadcasters who saw "fee-vee" as a threat to their advertiser-supported "free-vee." In 1955, however, the FCC started proceedings to determine whether pay-TV should be authorized. The broadcast entrepreneurs continued but the experiments failed. Pay TV via cable, however, was begun in 1975 by HOME BOX OFFICE (HBO) when that network began transmitting to cable systems via SATELLITE. Interest in pay TV via broadcasting languished, but was revived when the FCC rescinded all pay-TV rules related to broadcasting in 1982, opening up a short period of growth for what was now beginning to be called subscription television (STV).

As a form of pay TV, STV was to prosper for a few years in the 1980s, but it eventually failed because it could offer only one channel in what had become a multichannel world. Other forms of pay TV, MULTICHANNEL MULTIPOINT DISTRIBUTION SERVICE (MMDS) and some LOW POWER TELEVISION (LPTV) stations that scramble their signals, continue to seek subscribers in the 1990s. Cable is, however, the now-dominant form of pay TV. (See also COMMUNITY ANTENNA TELEVISION [CATV].)

pay-per-transaction   (PPT)   a method of home video retailing that allows the retail stores to lease prerecorded titles instead of buying them outright. They can therefore stock more copies of a popular "A" TITLE to satisfy customer demand. The stores divide the rental income from their customers with the leasing company that, in turn, shares it with the PROGRAM SUPPLIERS of the titles.

PBS   See PUBLIC BROADCASTING SERVICE (PBS).

PBS Program Fund the national program plan for acquiring programming for the PUBLIC BROADCASTING SERVICE (PBS) schedule. In 1991 it replaced the cumbersome but democratic Station Program Cooperative (SPC) that had served the PUBLIC TELEVISION (PTV) stations for nearly 20 years. The initial PBS Program Fund totaled some $100 million, with $22.5 million coming from the CORPORATION FOR PUBLIC BROADCASTING (CPB) and the rest from PBS and the stations. The stations continue to support the fund by paying program fees for the entire national program service (rather than individual programs), based on the formulas used to set PBS membership dues. The CPB continues to support the PBS Program Fund but retains some funds for its own program fund.

PCS initials standing for personal communications service. The term refers to the technology that could eventually become the primary form of two-way voice and data communications, replacing the telephone and computer. Such services will use a form of low-power digital mobile radio (similar to today's cellular radio technology) that (when used with SATELLITES or a cable FIBER OPTICS system and the routing capabilities of cable television) could revolutionize voice and data communications. Such systems could eventually be used in miniature telephone/computer terminals.

PEG channels an acronym for public access, eductional, and government channels in a cable system.

The phrase is derived from Section 611 of the CABLE COMMUNICATIONS POLICY ACT OF 1984, which permits a local FRANCHISE authority to "require as part of a franchise . . . that [some] channel capacity be designated for public, educational, or governmental use." The cable operators, however, have no editorial or program control over the channels.

Many school systems, libraries, or other governmental agencies have been granted control of some channels by the local cable system. Public-access channels are a somewhat different matter. Such channels are usually seen as a means of providing local residents with an electronic soapbox. They are offered free of charge, largely on a first-come, first-served basis, to individuals or groups. Although PEG and CUPU LEASED ACCESS CHANNELS are usually used to transmit local programs, they are both distinct from LOCAL ORIGINATION CHANNELS in that the control of the content of a program normally is not under the jurisdiction of the cable system. The final liability for content control of the access channels, however, continues to be a legal issue.

penetration the level or degree to which a specific medium reaches an audience or the extent to which a program or product is seen or purchased in a particular market. It is usually expressed as a percentage or a number. In home video, the videocassette recorder (VCR) may be said to have reached a penetration of 70 percent of the television homes. In broadcast television, penetration is expressed in RATINGS or SHARES. In cable television, penetration means the percentage of HOMES PASSED that are signed up as subscribers for cable services.

people meter a remote-controlled device connected to a television set that measures viewers' preferences and viewing times. It has buttons for each member of the family or visitors to the household and when the buttons are pressed, information (about who is watching what program and when) is noted. The data are stored in an in-home metering system until retrieved daily by the A. C. NIELSEN COMPANY computers at its processing center in Florida. (See also PEOPLE USING TELEVISION [PUT].)

people using television (PUT) an audience measurement that represents the percentage of people or the total number of people in a viewing area that are watching television at any given time. PUT levels refer to the viewing in a market as a whole rather than for individual stations, networks, or programs.

PUT measurements differ from the older HOUSEHOLDS USING TELEVISION (HUT) levels inasmuch as they calculate people rather than households. This audience research methodology was made practical and feasible in the late 1980s with the introduction of PEOPLE METERS.

Pepper Paper a FEDERAL COMMUNICATIONS COMMISSION (FCC) working paper issued in July of 1991. Formally titled *Broadcast Television in a Multichannel Marketplace,* it was written by staff members of the Office of Plans and Policy (OPP) and was popularly named after the head of the OPP, Robert Pepper. According to the study, "the broadcast industry has suffered an irreversible long-term decline in audience and revenue shares, which will continue through the current decade." The major villain is cable, according to the report. It will continue to affect all but the healthiest AFFILIATED STATIONS and strong INDEPENDENT STATIONS in the major MARKETS. The paper (which predicted events only up to 1999) was used to support the varied positions of the many aspects of the communications industry.

per inquiry (PI) an advertising practice in which payment is made to a television or cable operation based on a percentage of the money received by the advertiser through the sales generated by the COMMERCIAL on that medium. In its most common form, the operation transmits the commercial for a product that contains an 800 phone number or a post office box number. Rather than being paid for the COMMERCIAL TIME, the television or cable company receives a specific percentage of the monetary value of each unit ordered in response to the SPOT.

Another form of PI is based on the use of a 900 phone number. In this practice the viewer pays for the call, with the charges ranging from $.25 to $.75 per call. The income is shared among the local telephone company, AT&T, the service provider, and the network or station.

performance standards the minimum technical norms established by a city or county for cable companies operating under a FRANCHISE in its area. The engineering standards vary according to the current operating agreement and are usually renegotiated at renewal time, but the cable company must maintain those standards throughout the course of the contract.

persistence of vision the phenomenon of the brain and the eyes that makes possible the viewing of motion picture film and television. It occurs because the eye retains what it sees for only a short time. When a series of even slightly varying still images (motion picture frames) are seen at a speed faster than can be processed by the brain and the optic nerve connecting the eyes and the brain, the distinction between the static pictures is lost and movement is perceived to occur. A minimum rate of 10 images per second is usually required for the illusion of motion.

personal attack rules FEDERAL COMMUNICATIONS COMMISSION (FCC) rules that are a part of the FAIRNESS DOCTRINE. Along with the POLITICAL EDITORIAL RULES, they are designed to allow people who have been attacked on the air an opportunity to respond.

A "personal attack" is defined quite specifically. Criticism of a person's intelligence or ability is not a personal attack under this rule. Maintaining that a legislator is ignorant is not a personal attack but saying that

he has taken a bribe is. The attack must take place during a discussion of a "continuing issue of public importance."

Personal attacks on foreign groups or foreign public figures and personal attacks by legally qualified political candidates or their spokespersons in campaigns are exempted from the rule. Attacks occurring during news interviews or on-the-spot coverage of news or even during news analysis or commentaries are also exempted.

The personal-attack rules have been subject to a number of petitions to the FCC and court cases over the years. Their constitutionality, however, was upheld by the Supreme Court in the RED LION case.

Phi Delta Epsilon See SOCIETY FOR COLLEGIATE JOURNALISTS.

picturephone See VIDEOPHONICS.

pie formula a system of network programming in which the FORMAT (the crust) remains the same from week to week but the content (the filling) is different. The strategy has two purposes. Pie formula programming is economical in that sets and other aspects of production are unchanged, and audience loyalty is encouraged by setting the familiar in an unfamiliar or slightly different situation each week. SITCOMS and GAME SHOWS are classic examples of the pie formula.

piggyback commercials an advertising technique involving the scheduling of two COMMERCIALS for different products from the same company back-to-back. The commercials are very short and are designed to gain the maximum exposure for a company's products without increasing the amount of COMMERCIAL TIME that must be purchased.

pilot program a sample program of a proposed series, produced to introduce local television stations, cable or broadcast NETWORKS, and prospective sponsors to the potential of the new series. A pilot is sometimes created as the first episode of a series. A SITCOM pilot contains all of the elements of the projected series including the characters and their relationships to one another, the location, and the circumstances of the basic plot. A GAME SHOW pilot is shot with the hosts and participants in the studio setting actually playing the game. A variety or musical pilot is usually produced as a SPECIAL, designed to be aired even if the proposed series isn't sold. In both first-run syndication and network television, only one in five pilots is ever made into a series.

piracy the unauthorized use of COPYRIGHTED material. There are two types of piracy in the electronic communications world, signal theft and replication. The largest unauthorized use occurs in three areas: cable systems, SATELLITE-delivered program networks (signal theft), and home video (replication). Section 633 of the CABLE COMMUNICATIONS POLICY ACT OF 1984 prohibits pirated cable reception. Cable companies

have also installed SCRAMBLING hardware in their systems to stop signal theft and many systems (after a REBUILD) use ADDRESSABLE CONVERTERS. These devices have become an effective barrier to piracy because unpaid signals coming back to the headend can be easily traced.

Piracy by owners of TELEVISION RECEIVE ONLY (TVRO) satellite gear is more complicated. Concerned that an increasing number of TVRO owners were viewing programs without paying for them, the cable networks successfully lobbied Congress, which further amended Section 605 of the COMMUNICATIONS ACT OF 1934 (in the Cable Act of 1984) to specifically prohibit signal theft from satellites. The cable networks also began scrambling their signals. To receive a satellite signal, the TVRO owner must now lease or buy a descrambler and pay fees for the programs, similar to the fees paid by cable subscribers. And although some LOW POWER TELEVISION (LPTV) stations that operate PAY-TV services also scramble their signals, illegal descramblers are often devised and installed to thwart those efforts.

The most widespread piracy in the new media, however, occurs in home video because videocassettes are so easy to duplicate. Congress passed a stronger law in 1982, establishing more severe penalties for convicted video pirates, and the Federal Bureau of Investigation (FBI) and local authorities have become more active in enforcing the law. Local and state statutes against the practice have also helped, as has the installation of a MACROVISION signal on many new "A" TITLES that effectively hampers DUBBING.

The right of OFF-THE-AIR RE-CORDING for later playback in one's own home was established in the BETAMAX case in 1976. This TIME SHIFTING is not considered piracy and although it has not been tested in the courts, most people believe that taping from a cable system channel is, by extension, also legal. Taping and then charging admission to view the program, however, is clearly a violation of the copyright law, and is therefore piracy.

pitch   the process by which new business is solicited from a prospective customer or CLIENT through a sales presentation. In television and cable operations, an account executive will make a sales pitch to an ADVERTISING AGENCY to sell COMMERCIAL TIME. In CONSUMER ELECTRONICS, a sales clerk will describe the features of an electronic device and the benefits the customer will receive if it is purchased. Advertising agency personnel make presentations to prospective clients in an effort to persuade them to allow the agency to handle their ACCOUNT. (See also ADVERTISING AGENCY REVIEW.)

plant   the physical components of a cable system. The term is applied specifically to the HEADEND equipment but it is sometimes used to refer to all technical aspects of the operation, including TRUNK LINES, FEEDER CABLES, DROP LINES, AMPLIFIERS, and all other electronic gear.

The equivalent of a plant in the television broadcasting industry is FACILITIES.

pledge week   the two to four weeks during the year when special programs are telecast and the viewer is encouraged to make a donation and become a member of a PUBLIC TELEVISION (PTV) station. Membership costs vary and premiums (goods such as umbrellas, tote bags, and books) are often given with the membership to encourage viewers to donate funds. Pledge weeks are sometimes called "begathons."

plug   the process of promoting a product or service on a television program. Guests on a TALK SHOW often discuss their recent books or new movies in a seemingly casual way and thus "plug" the product. This form of advertising differs from a COMMERCIAL in that the personalities or the companies they represent are not charged a fee by the station, network, or cable system.

plumbicon tube   a pickup tube used in professional television cameras in broadcast and production operations. Sometimes called a lead oxide tube, it is expensive but it produces superior pictures. Three plumbicon tubes are required for each camera, one to pick up each of the primary colors.

pocketpiece (PP)   a small biweekly report that provides national RATINGS, SHARES, and some DEMOGRAPHIC information in a condensed form. It is published by A. C. NIELSEN. The booklets are small enough to fit into a coat pocket.

point of purchase   See POP.

pole rights   the right to attach hardware to telephone or light poles for the purpose of suspending COAXIAL CABLE. It is granted to cable system operators through an agreement with the local public utility companies. In the early days of the cable industry there were many legal battles over such rights. Some telephone companies refused to grant pole rights in the hope of curtailing cable growth. Congress finally addressed the problem in the Federal Pole Attachment Act of 1978, which gave the FEDERAL COMMUNICATIONS COMMISSION (FCC) jurisdiction to regulate the rates, terms, and conditions of cable pole attachments when they were not regulated by the state. Later the CABLE COMMUNICATIONS POLICY ACT OF 1984 required that states asserting jurisdiction publish regulations.

political editorial rules   FEDERAL COMMUNICATIONS COMMISSION (FCC) rules governing a station editorial that represents the official view of the licensee. The rules allow the person or persons affected by a station editorial a chance to state their side of the case on the air, in person or through a spokesperson. When a station endorses or opposes a legally qualified candidate or candidates, it must notify the other candidate(s) within 24 hours of the date and time of the editorial, send them a script or tape of it, and offer the candidate(s) or their spokespersons an opportunity to respond on the air. While similar to the EQUAL TIME (OPPORTUNITY) RULES in Section 315 of the COMMUNICATIONS ACT OF 1934, the political endorsement rules apply only to candidates who have been adversely affected by statements made by the licensee of the station and not by statements made by an opposing candidate(s). These rules do not, however, relieve the station of any obligation it may have under Section 315. (See also FAIRNESS DOCTRINE and PERSONAL ATTACKS.)

POP   initials that stand for "point of purchase" promotional materials in a retail setting. In home video stores, the materials consist of free-standing cardboard floor displays, sell sheets and brochures, TENT CARDS, banners, mobiles, and posters touting new title releases, or counter standups that promote a star. The POP materials are created by the PROGRAM SUPPLIERS and are distributed by them and their WHOLESALERS to encourage impulse sales on the premises of a video store.

pop-in   a short SPOT ANNOUNCEMENT on a cable or television operation that provides a brief burst of information not necessarily related to the advertiser's product or services. It is a type of IMAGE COMMERCIAL, designed to foster good will for the advertiser. Pop-ins are often scheduled during a holiday season when, for example, advertisers wish viewers a "Happy Thanksgiving."

portapack the early versions of portable cameras and videotape recording equipment, manufactured by SONY in 1970. The gear consisted of a small camera and a battery power pack that was worn around the waist, along with a half-inch reel-to-reel recording unit that was often carried in a backpack. The gear was heavy and cumbersome, however, and was eventually replaced by the smaller CAMCORDER units, but many people continue to call the modern portable assemblages by their original name. (See also ALTERNATIVE TELEVISION.)

positioning the attempt by ADVERTISING AGENCIES to place COMMERCIALS for CLIENTS' products or services in the most advantageous SPOTS on television or cable operations. A study of RATINGS and SHARES helps determine where to place a commercial in order to achieve maximum effectiveness.

The term is also used in home video, consumer electronics, and other commercial merchandising operations to define a strategy that is designed to distinguish a company from similar companies in the marketplace.

postproduction the procedure of compiling and editing portions of videotape recordings into a final polished program. Shots or bits of the program are assembled into sequences, sequences into sections, and sections into the finished program. The process is done in a videotape editing room or suite or at a commercial PRODUCTION FACILITIES COMPANY or postproduction firm (post house). During the postproduction process, personnel are said to be "posting" and after the job is completed, the program is said to have been "posted."

preamplifier a device that boosts an electronic signal prior to its further transmission. It is used to strengthen a weak signal before it is sent through the system. Its purpose is to bring the power of the signal up to a higher level where it will drive subsequent processors or other AMPLIFIERS in the system. It is known familiarly as a "preamp."

prebook the process in which a home video retail store informs a WHOLESALER of the number of units of a new title it intends to buy, prior to its official release (STREET DATE.) By establishing a specific booking date for preorders, the wholesaler and PROGRAM SUPPLIER get an idea of the demand for the new title.

preemptable rates special discount rates that are occasionally offered by account executives at television and cable operations to advertisers or ADVERTISING AGENCIES. The rates charged for the COMMERCIAL TIME are lower than those on the RATE CARD, but the time is sold on the condition that it will have to be relinquished if another advertiser offers to pay the regular fee.

preemption an action involving the temporary replacement of a program on a television station or cable system with another show. Stations sometimes preempt regularly scheduled programming to broadcast a program of unique local interest or a network may preempt a program to carry a presidential news conference or to report important news events, and then later resume its regular program schedule.

prepacks   a package containing more than one blank or prerecorded videocassette for sale to customers. The videocassettes are packaged by SHRINKWRAPPING or placed in special cardboard outer coverings or containers. Prepacks range from "twin packs" (two-packs) of prerecorded titles and "three-packs" or "four-packs" of blank videocassettes. They are offered at a discount that is less than the cost of the individual videocassettes.

presentation   in general, any speech or performance designed to elicit some form of response by an audience. In the advertising industry, it is used to describe the live, face-to-face delivery of an idea to a CLIENT by an ADVERTISING AGENCY. Some presentations are designed to attract new clients and are called a PITCH. The agency's purpose is to convince the potential client of its experience and abilities in order to win the ACCOUNT. Presentations are also made to existing clients to elicit their response to and approval of a campaign or a specific segment of some advertising plans. (See also ADVERTISING AGENCY REVIEW.)

pressure microphone   See OMNIDIRECTIONAL MICROPHONE.

Prestel   the British version of two-way VIDEOTEXT. Prestel (short for "**press** and **tell**") is operated by the British Postal Service and uses telephone lines connected to home, school, and business television sets and keyboards. The technology used is the same as for the British teletext systems CEEFAX and ORACLE.

prime time   according to the FEDERAL COMMUNICATIONS COMMISSION (FCC), the four broadcasting hours between 7:00 P.M. and 11:00 P.M. eastern and Pacific standard time, and 6:00 to 10:00 P.M. central and mountain standard time. (The FCC definition of prime time also includes PRIME-TIME ACCESS time.) A. C. NIELSEN does not follow that definition and considers only three hours— 8:00 to 11:00 P.M. EST and PST and 7:00 to 10:00 P.M. CST and MST— as prime time. Whether three or four hours, prime time is the most popular and highly viewed DAYPART in the schedule of a television station, network, or cable system.

prime-time access   the DAYPART in the program schedule of a television station or network that is the time period just before PRIME TIME. It is sometimes known as "prime access" or simply "access." In 1970 the FEDERAL COMMUNICATIONS COMMISSION (FCC) adopted the PRIME TIME ACCESS RULE (PTAR) that limited the amount of programming that a network could provide to its affiliates to three out of the four prime-time hours. The individual local stations normally broadcast local news programs from 7:00 to 7:30 P.M. (EST and PST), so the networks mutually agreed to turn back 7:30 P.M. to 8:00 P.M. (eastern standard time and Pacific standard time) Monday through Saturday to the local stations. This time became known as "prime time access." (The time affected in the central and mountain time zones is one hour earlier.) Since then, the time covered by the phrase has also been expanded by many SYNDICATORS, ADVERTISING AGENCIES, and networks to encompass the entire period between 6:00 P.M. and 8:00 P.M.

prime time access rule (PTAR) the FEDERAL COMMUNICATIONS COMMISSION (FCC) ruling that limits the number of PRIME-TIME hours a network can program to three hours each night. The ruling went into effect in 1971 and has been cited as the most important reason for the growth of SYNDICATION and INDEPENDENT STATIONS in the 1970s and 1980s.

The FCC confined the networks, under the PTAR, to providing material to their affiliated stations for only three of the four prime-time hours. The rule applied only to network-affiliated stations in the top 50 markets in the country. The most immediate effect of the rule was to open up a half-hour period for the scheduling of syndicated programs and to legitimatize the BARTER system. Faced with no network programs at 7:30 P.M., the affiliated stations in

the smaller 150 markets began to schedule off-network syndicated programs. And because the rule forbade the affiliated stations in the top 50 markets from scheduling off-network syndicated shows, they turned to FIRST-RUN SYNDICATED PROGRAMS. The rule prompted a resurgence of GAME SHOWS, which were stripped five nights per week by the stations.

The PTAR has been the subject of repeated controversy, and many appeals and petitions have been made to change some of its provisions. Revisions in 1974 allowed the networks to schedule news, documentaries, and children's programs at 7:30 P.M. and there continue to be attempts to eliminate the PTAR rules entirely.

PRISM a 24-hour PAY (PREMIUM) CABLE network headquartered in Bala Cynwyd, Pennsylvania. Some three-quarters of its offerings are first-run movies and the remainder are professional and college sporting events.

private cable see SATELLITE MASTER ANTENNA TELEVISION (SMATV).

private television See CORPORATE TELEVISION.

PRIZM an acronym standing for Potential Rating Index for Zip Markets. The CLUSTER ANALYSIS research methodology electronically sorts the 36,000 Zip codes in the United States into 40 lifestyle categories, or clusters, with descriptive labels such as "Fun and Station Wagons" and

"Gray Power." Each cluster is also organized into 12 broader social groups. Each group and cluster has a particular set of viewing and buying patterns and preferences.

Local television stations sometimes use PRIZM to assist in selecting programming and as a sales tool for potential advertisers. Cable systems sometimes use the method to identify demographic-specific (demo-specific) subscribers for SPOT advertising sales as well as potential subscribers. (See also ACORN.)

processing amplifier an electronic device used to clean up and improve a video signal. It is known familiarly as a "proc (pronounced prock) amp" and is sometimes also called a "stabilizer." A processing amplifier makes a video image more stable by removing JITTER. It can also increase the contrast in an image, vary the color, realign the brightness, and sharpen the synchronization (SYNC) of the signal. A proc amp is used between the source of the signal (such as a videocassette recorder [VCR] and its destination (another VCR or a MONITOR).

production facilities companies businesses that rent studios, REMOTE trucks, VIDEOTAPE EDITING equipment, and other gear used in the production of cable and television programs. They are often called "production houses" or "postproduction houses" (posthouses). All segments of the cable and television industries use such facilities on occasion to supplement their own equipment and services.

Some firms (called full-service companies) offer full production as well as POSTPRODUCTION services, including editing and DUBBING. Others specialize only in renting remote trucks and EFP equipment, and still others concentrate on film-to-tape transfers. Even within these subareas, some firms will rent very specific editing equipment or provide unique music or computer animation services. These quite specialized houses are known as "boutiques."

Professional Audio/Video Retailers Association (PARA) a nonprofit association of retail dealers in the more sophisticated audio/video equipment. The association holds seminars and training courses in sales and retail management, bestows awards, and publishes a newsletter.

Professional Film and Video Equipment Association (PFVEA) a nonprofit organization consisting of manufacturers of professional video and film equipment as well as repair personnel at retail outlets. The group's activities include publishing a "Missing Equipment List," "Equipment of Questionable Origin List," and a newsletter.

program analyzers devices used to check and record the reaction of an audience to a television program. Volunteers representing a typical audience are assembled in a room to watch a program or commercial. They are given a program analyzer and asked to push buttons indicating their acceptance or rejection of elements of the show at any time. In other circumstances, sensors gauge their physiological reactions to what they are seeing. The program analyzer records the group's second-by-second reaction to the show for later analysis by researchers.

program log a daily master plan of a television station that lists and schedules all operational elements in the broadcast day by the minute and second. It records the schedules of programs, COMMERCIALS, and PUBLIC SERVICE ANNOUNCEMENTS (PSA). Often simply called "the log," it is the single most important document in a station, itemizing the details of the operation from SIGN-ON to SIGN-OFF. The program log is often generated by a computer and the entries of the operational personnel are also entered into a computer for later analysis. At the end of the day a complete record of every activity at the station has been recorded and documented in much the same way as a ship's log. Until 1981 the FEDERAL COMMUNICATIONS COMMISSION (FCC) required all stations to keep a daily program log. In that year, the Commission dropped that rule.

program supplier a company that develops and produces or acquires the home video rights to a prerecorded video title, then markets and sells the title to the trade. The companies create and promote LABELS bearing their own or related names for use in their sales efforts. The term also identifies an individual who produces and markets prerecorded video programming.

projection television units electronic devices that throw a television image on a large screen for optimum viewing by large groups of people. The units are increasingly common as a part of a home theater environment. The largest projection devices are commercial units such as the JUMBOTRON, which can project an image of $23' \times 32'$ in an outdoor setting. Smaller units are used in auditoriums and arenas. In general, the larger the screen, the less brightness and clarity the picture has. Front projection devices require a special screen of from $4'$ to $25'$ (measured diagonally) placed several feet away from the projector. Rear projection sets are self-contained units that throw an image onto a large 40-inch to 70-inch diagonally measured screen.

Promax See BROADCAST PROMOTION AND MARKETING EXECUTIVES INC. (BPME).

promos verbal shorthand for "promotional announcements," short audio, film, or videotape pieces that give the audience information about upcoming programs. On-air promos are prerecorded SPOT ANNOUNCEMENTS, designed to interest and attract viewers to future individual programs or to entice them to tune into a new series. Such promos feature clips from the show and inform the viewer when it will be aired. The term is also used for the verbal endorsements delivered by sportscasters near the end of sporting events and by newscasters urging the audience to "stay tuned for . . ." Promos are the electronic version of TUNE-IN ADVERTISING. (See also CONTINUITY.)

proof of performance　a written notice, sent to the FEDERAL COMMUNICATIONS COMMISSION (FCC) by an applicant for an FCC LICENSE for a station. It certifies the satisfactory performance of the TRANSMITTER and ANTENNA of the new or renovated broadcast station. After the applicant receives a CONSTRUCTION PERMIT (CP) from the FCC and builds the facility, engineering tests and measurements are conducted to determine if the technical parameters contained in the application have been met. Successful proof of performance of the electronic equipment is required before the FCC will issue a license to broadcast.

prosumers　consumers who seek to use video equipment in a professional manner or professionals who purchase high-end consumer video equipment for use in their small business operations. Many manufacturers offer equipment with professional video production features at prices slightly higher than consumer gear. Their prosumer CAMCORDERS, small VIDEOTAPE FORMATS, sound and lighting equipment, and accessories straddle the fence between the low end of industrial and professional lines and high-end consumer equipment.

psychographics　the classification of television viewers by their values and life-styles, which are analyzed before creating an advertising CAMPAIGN. The motivation of consumers, their buying habits, preferences in colors or packaging, purchase behavior, and other psychological data are gathered in psychographic research. Psychographics provide a more in-depth profile of consumers than DEMOGRAPHIC analysis, which relies on such primary characteristics as age, sex, and income.

public access channels　See MIDWEST VIDEO II DECISION and PEG CHANNELS.

Public Affairs Video Archives　an archive that preserves, catalogs, and distributes all programming on both channels of C-SPAN (the Cable Satellite Public Affairs Network). It operates independently and exclusively for educational purposes with the co-operation of C-Span. The organization has recorded and cataloged all C-Span programming that has been telecast since October 1987.

Public Broadcasting Act　a 1967 law that rejuvenated the EDUCATIONAL TELEVISION (ETV) industry and moved it in a different philosophical direction with a new name, PUBLIC TELEVISION (PTV). A report from the CARNEGIE COMMISSION ON EDUCATIONAL TELEVISION (CARNEGIE I), which was released in 1967, was the basis for the law. The Act addressed three issues in three parts: (1) the need for the construction of stations, (2) the establishment of a nonprofit national corporation to lead the movement, and (3) a call for a study of INSTRUCTIONAL TELEVISION (ITV). The first section extended the life of the EDUCATIONAL TELEVISION FACILITIES ACT for three years and authorized another $38 million in matching funds for the construction and improvement of stations.

The second part of the Act created the nonprofit CORPORATION FOR PUBLIC BROADCASTING (CPB), which led to the later establishment (with CPB funds) of the PUBLIC BROADCASTING SERVICE (PBS) and National Public Radio (NPR). Part three authorized the Department of Health, Education, and Welfare (DHEW) to conduct a study of instructional television and radio, which was subsequently completed by the COMMISSION ON INSTRUCTIONAL TECHNOLOGY (CIT).

On November 7, 1967 President Johnson signed the bill into law, and noncommercial broadcasting had a new name.

Public Broadcasting Financing Act of 1975 (PBFA) major legislation in the development of PUBLIC TELEVISION (PTV), passed as an amendment to the COMMUNICATIONS ACT OF 1934. The 1975 Financing Act authorized federal funds for the PTV system for five years and made actual advance appropriations for two years. It made "forward funding" possible and protected the system from retaliatory budget cuts due to controversial programs. However, the longer authorization period was short-lived. The PUBLIC TELECOMMUNICATIONS ACT OF 1978 reduced the authorization term for the CORPORATION FOR PUBLIC BROADCASTING (CPB) to a three-year period but maintained the two-year appropriation cycle.

Public Broadcasting Service (PBS) the national PUBLIC TELEVISION (PTV) programming and interconnection service. Established in 1969, PBS became a national membership organization of PTV stations in 1973, representing the interests of its member stations to federal agencies, Congress, the CORPORATION FOR PUBLIC BROADCASTING (CPB), and the public. In 1979, as part of an industry reorganization, PBS restructured itself to focus only on the design and delivery of programming. The National Association of Public Television Stations (NAPTS) (now the ASSOCIATION FOR PUBLIC BROADCASTING [APB]) assumed the lobbying and representation functions.

PBS is not a network in the strictest sense, but is a private, nonprofit membership organization that operates a national satellite interconnection system (inaugurated in 1978) that links all of the nation's PTV stations.

PBS acquires and distributes programming but, unlike its commercial network counterparts, it does not produce programming. Each local PTV station pays PBS to receive programming and other services. PBS also operates an ADULT LEARNING SERVICES division, which transmits TELECOURSES for DISTANCE EDUCATION purposes.

The PBS organization is responsible to a Board of Directors composed of professional managers and lay leaders from the member stations. It is funded by the CPB and membership dues.

public domain   the status of a work that has never been copyrighted or a work on which the COPYRIGHT has expired. Works in the public domain, including motion pictures, television programs, and commercials, can be used by anyone at any time without permission from the creator and at no cost.

public file   a file of documents that must be available for public inspection at any radio or television station, according to FEDERAL COMMUNICATIONS COMMISSION (FCC) rules. This public file must include Ownership Reports, the station's EQUAL EMPLOYMENT OPPORTUNITY (EEO) plan and Annual Reports to the FCC, a record of all political broadcasts and political time requests, letters from the public, and a copy of every application to the Commission. In addition, a list of programs that the station has broadcast about significant community issues in the previous three months and any complaints made by the public to the FCC about programming must be included in the file.

public notice   a printed statement in the classified section of a local newspaper or an announcement over the air about the proposed actions by a broadcast licensee. All applicants for a CONSTRUCTION PERMIT (CP) for a new broadcast station or for any modification in an existing application before the FEDERAL COMMUNICATIONS COMMISSION (FCC) must give public notice of their actions. They must inform people about any FCC responses concerning their applications including any COMPARATIVE HEARINGS that are scheduled by the Commission. Holders of an FCC LICENSE must also inform the public about any changes or modifications in their license and provide information concerning their applications for a LICENSE RENEWAL. The purpose of the notice is to give the public an opportunity to comment on the applications.

public performance of copyrighted video material   the concept and issue of what constitutes a public performance of copyrighted video material. Videotaped programs that are purchased, rented, or copied from OFF-AIR VIDEO RECORDING and played back at assemblies, children's story hours, or club meetings are the kinds of "performance" in question.

The COPYRIGHT ACT OF 1976 strengthened the rights of the creators of programs over their performance and display, but it also exempted certain educational performances from the need to obtain permission or pay fees. The EDUCATIONAL USE OF COPYRIGHTED VIDEO MATERIAL exempts playbacks in face-to-face instruction in a nonprofit educational institution.

The Act, however, gave the copyright owner the exclusive right "to perform the work publicly." In Section I of the Act, "publicly" was defined as a performance taking place anywhere "open to the public or at any place where a substantial number of persons outside a normal circle of a family and its social acquaintances is gathered." In such a circumstance, permission and a license from and fee to the copyright holder is usually required before the showing of a videocassette.

Some interpretations of "publicly" in the law expanded it to apply to viewing by a single library patron in a carrel so that such a playback would be prohibited. The American Library Association (ALA) has disagreed. The ALA contends that as long as a showing is limited to a family or one individual, it is not a public performance. Some of the proponents of such "in-house" use also believe that such viewing is permitted as part of the FAIR USE DOCTRINE. The ultimate resolution of the "public performance" issue will come in the courts.

public service announcement (PSA) a short, on-the-air announcement that resembles a COMMERCIAL. A PSA, however, promotes a cause or noncommercial service in the public interest. Sometimes called by an old radio term, "sustaining announcements," PSAs are usually developed by nonprofit organizations or government agencies and are transmitted by a television station, network or cable system at no charge to the organization. They focus on an idea or concept and are sometimes created as a *pro bono* effort by the ADVERTISING COUNCIL INC. Red Cross and United Way PSAs are typical of the GENRE.

Public Telecommunications Financial Management Association (PTFMA) a nonprofit group consisting of public broadcasting organizations that seek to foster new financial and accounting techniques in the noncommercial industry. The organization provides a venue for members to exchange ideas.

public television (PTV) the noncommercial segment of the television industry in the United States. Although the term has been in use since 1967, this type of television is still officially labeled "noncommercial EDUCATIONAL TELEVISION (ETV)" by the FEDERAL COMMUNICATIONS COMMISSION (FCC). To many U.S. citizens, noncommercial television's fundamental purpose is to present INSTRUCTIONAL TELEVISION (ITV) programming and TELECOURSES to facilitate DISTANCE EDUCATION. To others, its purpose is to provide programming for specific minorities and children through a service that is controlled by the public. And to others, the PTV system is viewed as a necessary alternative to commercial television where high-quality programming—including dance, opera, ballet, public affairs, and drama—is presented. PTV fulfills all three functions.

Noncommercial television began in 1952 with the reservation of channels by the FCC for noncommercial use. The FCC's SIXTH REPORT AND ORDER set aside 242 channels for the specific purpose of "serving the educational needs of the community." In 1967, the CARNEGIE COMMISSION ON EDUCATIONAL TELEVISION: (CARNEGIE I) completed a report that proposed major changes in the objectives and funding of noncommercial television. The Carnegie Commission used the term "public television" to distinguish the new concept from what was regarded by some as the limited image of the term "educational television."

The Commission's recommendations were incorporated into federal legislation, which culminated in the PUBLIC BROADCASTING ACT of 1967. The new act created the CORPORATION FOR PUBLIC BROADCASTING (CPB) to assist the local stations in the full development of the industry. Although each local station operates independently, they all form the basis for the PTV system in the United States.

The stations cooperate extensively, holding memberships in REGIONAL TELEVISION NETWORKS, the PUBLIC BROADCASTING SERVICE (PBS) (which provides national programming), and the ASSOCIATION FOR PUBLIC BROADCASTING (APB) (which represents them before Congress).

pushing the envelope   the stretching of the contents of a program or commercial to the limits of propriety and good taste. Pushing the envelope is achieved by incorporating new or different ideas without offending the audience. The phrase was originally coined by U.S. test pilots who sought to break the sound barrier. (See also TABLOID TV PROGRAMMING.)

Q

quadruplex (quad) videotape recording   a videotape recording system developed by the Ampex Corporation in 1956. It effectively solved the problem of recording enough information on magnetic tape to ensure a good image and sound, without using miles of tape and wearing out the heads. Four heads (the quad in quadruplex) were mounted on a rapidly spinning drum, which rotated across the width of the 2-inch videotape. The tape itself moved through the mechanism at the relatively slow rate of 15 inches per second (IPS). The combination of the rapidly spinning heads and the slower moving tape resulted in an effective speed of 1,500 inches per second. This was more than enough to create an excellent black-and-white picture. By 1958 a color version had been developed and was marketed.

The 2-inch quad machines became the standard in the professional television industry during the 1960s and 1970s but the gradual improvement in the quality of HELICALSCAN VIDEOTAPE RECORDING machines in the 1980s reduced their dominance. The smaller 1-inch reel-to-reel helical machines gradually replaced the old quads.

quarter-hour audience (AQH)   the number of people watching a television station in a specific 15-minute period, as measured by research organizations. To be a part of such an audience, the viewer must be watching at least five consecutive minutes during that quarter-hour period.

QUBE   an early version of INTER-ACTIVE TELEVISION developed by Warner Communications and inaugurated in the company's Columbus, Ohio cable system in 1977. The basic element of QUBE was a small box with response buttons at the cable subscriber's home, which allowed the user to send an electronic signal back to the HEADEND of the system. The viewer could respond to a question or to a written query on the screen by pushing a button. A computer instantly analyzed the responses of all of the participating viewers. QUBE attracted widespread media interest. Versions of it were installed in Warner's other cable systems in Pittsburgh and Dallas. Initially the customers were offered the various services free of charge, but later they were required to pay for them. After the novelty wore off, there was a decline in use. After years of consistent losses, Warner-Amex disbanded the operation in 1984. (See also VIDEOWAY.)

quicksilver scheduling   a program scheduling practice in which less-than-popular programs are replaced or PREEMPTED quickly, often without sufficient notice to the viewing audience. Today, the networks are quick to cancel low-rated shows and replace them with others or move a program from one evening slot to another night and time. Such scheduling patterns make the search for a specific program as elusive as mercury (quicksilver) and fosters GRAZING and ZAPPING.

quintile   any of five equal groups within a measurement. One-fifth, for example, can represent the magnitude of television viewing in an audience measurement system such as the heaviest (or lightest) viewing quintile within the sample.

QVC Network Inc.   a cable shopping channel offering merchandise and products for cable subscribers at home. In 1994, the company engaged in a major fight with VIACOM ENTERPRISES to acquire PARAMOUNT COMMUNICATIONS INC.

R

rack jobber   a type of video wholesaler that sets up and maintains a display or rack of prerecorded videocassette titles in a retail outlet that normally would not carry such an item. The operations are often established in convenience stores or drug and grocery stores and are stocked and periodically restocked by the wholesaler, who pays a percentage of the sale of each title to the store owner.

rack mounting metal racks to which electronic equipment is attached in television control or editing rooms. The ELECTRONIC INDUSTRIES ASSOCIATION (EIA) devised a standard size for the racks to enable most electronic gear to fit into them. The racks are slightly more than 19 inches wide, and all engineering equipment of that size (or smaller) can be easily placed in them. Electronic units that are designed to be rack-mounted have screw-mounts so they can be secured to the racks.

Radio and Television Research Council (RTRC) an organization that involves professionals engaged in radio and television research. It seeks to improve techniques in audience research methods by lectures and discussions at meetings held the third Monday of each month in New York City.

Radio Corporation of America (RCA) a diversified electronics and communications company founded by the General Electric Company (GE) in the fall of 1919. It became one of the most powerful entities in broadcasting history. In 1932 GE was forced by the federal government to sell its shares in the company because of antitrust considerations. RCA became a leader in the manufacture of broadcast equipment and radio receivers and created the NATIONAL BROADCASTING COMPANY (NBC) in 1926, the first organization developed to operate a radio network.

RCA began experimenting with television in 1931 and demonstrated the miracle at the 1939 World's Fair in New York. The company established the NBC television network after World War II and was a leader in the manufacture of TV sets. A system based on RCA's electronic color television inventions was adopted as the standard for the United States by the FEDERAL COMMUNICATIONS COMMISSION (FCC) in 1953 and RCA's NBC subsidiary pioneered in offering programs in color. The firm manufactured television cameras, transmitters, a QUADRUPLEX (QUAD) VIDEOTAPE RECORDING machine, and other professional equipment.

The company's fortunes declined in the 1970s and 1980s, however, with losses suffered from the development of the CAPACITANCE ELECTRONIC DISC (CED) and the discontinuance of its television set manufacturing. RCA was sold to GE, its original parent, in 1985.

Radio Television News Directors Association (RTNDA) a nonprofit organization of news chiefs at radio and television stations and cable organizations. The group seeks to improve electronic journalism and to encourage education in journalism.

Rainbow Program Enterprises a subsidiary of CABLEVISION SYSTEMS CORPORATION that owns and operates a number of PAY (PREMIUM) CABLE as well as BASIC CABLE networks. The pay services include BRAVO and American Movie Classics, while SPORTSCHANNEL AMERICA and its 10 regional affiliates and News 12 Long Island are advertiser-supported.

raster   the pattern of viewable horizontal SCANNING LINES that form the image on a television screen. It is comprised of the scanned, visible portion of a CATHODE RAY TUBE (CRT). While the FIELD and FRAME are the entire image, the term raster is used to describe that portion that is visible within the parameters of a MONITOR.

rate card   a small booklet or brochure that lists the costs of COMMERCIAL TIME on a television or cable operation. It details the charges for specific times and any rules or restrictions related to advertising on that particular medium. It also indicates various discount plans including PRE-EMPTABLE RATES.

rates   charges made by cable or television operations for COMMERCIAL TIME. They are based on the RATINGS and SHARES achieved by various programs. The rates are usually spelled out in the medium's RATE CARD, but sales of SPOT time are often negotiated or made "off the card." (See also END RATES and PRE-EMPTABLE RATES.)

ratings   estimates of the size of a television or cable audience compared to the potential audience. They are compiled by audience research companies such as A. C. NIELSEN and are indicative of the relative popularity of a given program compared to other programs. ADVERTISING AGENCIES use them to determine where a CLIENT's COMMERCIALS should be placed.

The ratings are compiled by col-lecting data from PEOPLE METERS and by the use of the DIARY SYSTEM in the more than 200 DESIGNATED MARKET AREAS (DMAs) and AREAS OF DOMINANT INFLUENCE (ADIs) in the United States. The combined information measures the number of TV sets actually tuned to a program as compared to the total number of TV sets in that area. The rating is expressed in terms of a percentage or a point. One rating point represents 1 percent of all of the households in the MARKET. Ratings can be projected to estimate the size of the national audience by applying the percentages to the number of television sets in the United States. In addition to the audience size, the ratings systems also develop data on DEMOGRAPHICS, including the sex and age of the viewers.

Local ratings are collected for given periods, usually four times each year (called SWEEPS), although Nielsen also provides weekly ratings of network shows and, on request, also supplies OVERNIGHT RATINGS (the "overnights") for network programs. The ratings are published periodically after the sweeps in booklet form (called a BOOK) or in POCKETPIECES. (See also HOUSEHOLDS USING TELEVISION (HUT), PEOPLE USING TELEVISION (PUT), and SHARES.)

reach   See CUME.

reaction shot   a camera shot that shows the response to a speech, performance, or other action contained in the ongoing scene. The shot can show anger (at a remark), joy (at some news), or surprise (at a question). The ultimate reaction shot is a standing ovation. (See also FRAMING and OVER-THE-SHOULDER SHOT.)

reality commercials television commercials that mimic real-life situations. They appeared infrequently in the 1960s and 1970s but became more prevalent in the 1980s. Simulating a home video production or an ALTERNATIVE TELEVISION piece, the COMMERCIALS are often shot on grainy film with shaky hand-held cameras in a *cinema vérité* style. The actors resemble people next door rather than glamorous personages and the whole effect is an attempt to persuade the viewer to associate the SPOTS with the realism of a DOCUMENTARY, thereby lending credibility to the product.

reality programming programming that relies on soft but actual news activities or events. Based on "reality," as opposed to the fiction of SITCOMS, the programming features real people in real-life situations in pseudo MAGAZINE FORMAT (PROGRAMMING). The circumstances, however, are usually sensational, outrageous, or scandalous. "Lifestyles of the Rich and Famous" and "America's Funniest Home Videos" are examples of reality programming. (See also TABLOID TV PROGRAMS.)

rear screen projection a production technique in which a FILM LOOP, FILM CLIP, or videotape is projected onto a translucent screen to create a still or moving background for a set.

Actors or other talent perform in front of it. Slides of any size can be projected onto a rear screen to provide a static background or to display charts or graphics in an INSTRUCTIONAL TELEVISION program. Rear screen projection is also often used on newscasts to introduce live or taped reports from a remote location.

rebuild the physical improvements made in a cable system by the replacement of various electronic components and wiring. In the process, power supplies, AMPLIFIERS and other electronic gear are replaced with state-of-the-art technology, and COAXIAL CABLES (along with FEEDER and DROP LINES) are replaced with a FIBER OPTIC system to increase channel capacity. (See also NEW BUILD and UPGRADE.)

Red Books A pair of advertising directories with red covers published by the STANDARD RATE AND DATA SERVICE (SRDS). *The Standard Directory of Advertisers* lists more than 17,000 companies that advertise nationally along with their ADVERTISING AGENCIES and their budgets devoted to advertising. A companion book, *The Standard Directory of Advertising Agencies,* lists all of the agencies by state and their addresses, phone numbers, and executives. Both books are published annually and are very occasionally referred to by their official name, *The Standard Advertising Register.*

Red Channels a small book, subtitled "The Report of Communist Influence in Radio and Television." It was published in 1950 by an independent watchdog group called the American Business Consultants. The book contained some 200 pages of profiles of 151 broadcast personalities who were suspected of being Communists or having Communist sympathies. It immediately became a source for BLACKLISTING and ushered in a reprehensible era in broadcasting.

Red Lion case a Supreme Court decision in 1967 that upheld the constitutionality of the FAIRNESS DOCTRINE of the FEDERAL COMMUNICATIONS COMMISSION (FCC) and the COMMUNICATIONS ACT OF 1934. It is viewed as a landmark case in determining broadcasting's relationship to the First Amendment. The decision implied that broadcasting must be subject to different freedom-of-speech standards than the press because of the physics of scarcity (the limited number of FREQUENCIES). It also affirmed the obligation of the licensee of a station to present different viewpoints and, in effect, maintained that the listener's right to information supersedes the broadcaster's right of free speech.

reformatting redesigning broadcast programs for another medium. A television talk or interview show can be adapted to or combined with dramatic, musical, or other elements to create a "new" program for release on cable or home video. In most instances the changes are relatively minor and often consist of the insertion of one or two small sections

and some different transitions between elements of the show.

regional television networks permanent regional networks that are an integral part of the nation's PUBLIC TELEVISION (PTV) system. There are four regional PTV networks. Each covers a particular geographic area and provides programming and other services to member stations. Each of the four networks has a satellite UPLINK, acquires and distributes programs, operates INSTRUCTIONAL TELEVISION program services for member stations, and develops CO-PRODUCTION ventures. (See also CENTRAL EDUCATIONAL NETWORK [CEN], EASTERN EDUCATIONAL NETWORK [EEN], GROUP BUYS, INTERREGIONAL PROGRAM SERVICE [IPS], PACIFIC MOUNTAIN NETWORK [PMN], and SOUTHERN EDUCATIONAL COMMUNICATIONS ASSOCIATION [SECA].)

release the number of times a program may be broadcast. A station buys a license from a SYNDICATOR to air a program a certain number of times and each play is considered a release. The term is also used as a verb to indicate the placement of a program in distribution. Movies are released for PAY (PREMIUM) CABLE or home video use by the Hollywood studios and programs are released in OFF-NETWORK syndication by distributors.

religious programming television or radio programming that demonstrates a system of beliefs or promotes the practice of extolling a superhuman or divine power, usually including a philosophy and/or a code of ethics. In early television, there were three basic types of religious programming: discussions or interviews related to spiritual or ethical issues; syndicated dramas that revolved around a religious theme and actual worship services, usually REMOTE pickups from churches or synagogues. They were produced by mostly mainstream denominations and aired free of charge by broadcasters. In the 1960s and 1970s, however, some Christian evangelical groups began to buy time on television stations and cable systems in order to proselytize for their faith. Stations dropped the sustaining (free) shows for the income from the evangelists.

As the cable industry and satellite interconnection developed, the fundamentalist broadcasters embraced the new medium and at one time a number of cable networks were devoted exclusively to religion. The cable networks and fundamentalist programming reached their peak in the 1980s before declining as a result of sexual and financial scandals among the evangelicals.

(See also AMERICAN FAMILY ASSOCIATION, ASSOCIATION OF CATHOLIC TV AND RADIO SYNDICATORS, CATHOLIC BROADCASTERS ASSOCIATION, INTERNATIONAL CATHOLIC ASSOCIATION OF RADIO AND TELEVISION, JESUITS IN COMMUNICATION, NATIONAL RELIGIOUS BROADCASTERS, and UNDA–USA.)

remote programs produced or recorded live at a distance from the television studio. They contain all of the real-life atmosphere that an on-location setting can provide. Separated from the confines and artificiality of a studio, both crew and performers often are more informal and spontaneous. The subject matter of remotes varies from sports to parades to the coverage of national events.

rental cycle (video cassettes) the annual seasonal cycle for the rental of videocassettes from retail stores. With the exception of January, the first quarter of the year (January/February/March) is somewhat flat, but the second quarter (April/May/June) tends to be the weakest, with a steady decline after Mother's Day in May. The third quarter (July/August/September) is the second best of the year for videocassette rentals. The fourth quarter (October/November/December) is the best, with a spurt in rental activity occurring around the Thanksgiving holiday and during the Christmas and Hanukkah seasons.

reps See STATION REPRESENTATIVES.

reruns episodes from series or single programs repeated on the same station, network, or cable, MULTICHANNEL MULTIPOINT DISTRIBUTION SERVICE (MMDS), or LOWPOWER TELEVISION (LPTV) systems. Although the viewing audience tends to call all repeat programming "reruns," the term technically applies only to those programs originally aired on a station or system and repeated on that same operation. Programs in OFF-NET-

WORK syndication are therefore often not reruns, inasmuch as they are transmitted on different outlets.

residuals payments to creative personnel for the repeat use of their movies, programs, or COMMERCIALS. By individual or union contract, actors, performers, writers, and other creative personnel receive fees for the subsequent showings of their movies, programs and commercials. The rates are usually set by the AMERICAN FEDERATION OF TELEVISION AND RADIO ARTISTS (AFTRA) and the SCREEN ACTORS GUILD (SAG).

resolution   the relative quality of a picture on a CATHODE RAY TUBE (CRT). Although there are two types of resolution (horizontal and vertical), more attention is given to vertical resolution, which is usually expressed in terms of the number of SCANNING LINES per FRAME that can be seen when using test equipment. Vertical resolution is a measure of how sharply the electron beam can be focused on the face of the CRT. The smaller the dot that can be produced, the more scan lines can be placed on the screen, and the higher the resolution. Horizontal resolution is determined by how fast the video circuits can process the signal to vary the intensity of the dot.

Retransmission   Consent   Rules FEDERAL COMMUNICATIONS COMMISSION (FCC) rules that give broadcasters the right to negotiate with cable systems for the transmission of their signals. The regulations are coupled with MUST-CARRY RULES. Every three years, commercial television stations must choose whether to require carriage or negotiate a retransmission fee with local cable systems. PUBLIC TELEVISION (PTV) stations are exempt from the rule.

reverse angle shot   a camera shot that views the subject from an angle that is opposite to that used in the preceding shot. For example, a shot from the front of a person may be followed by a shot from the back. In live television, a full 360° shot is difficult because each camera will be in the other's shot. Most reverse angle shots on television are, therefore, limited to about 150°; many are OVER-THE-SHOULDER SHOTS. (See also FRAMING and REACTION SHOT.)

RF   an electromagnetic signal below the infrared but above the audio FREQUENCIES. It is short for radio frequency. In television broadcasting a modulator (or RF generator) combines the video and the SYNC signals (known as composite video) and the audio signal and impresses them onto a steady signal at a particular frequency or specific channel. The TRANSMITTER and ANTENNA broadcast on that channel and the home antenna picks up the RF signal. A tuner in the TV set tuned to that channel receives and demodulates the RF signal and separates the signal back into audio, video, and SYNC pulses, for viewing on the home screen.

ripple effect   a phenomenon occurring in videotape timecode AB ROLL EDITING. When the position or length of an edit is changed from the original master list, the starting times and positions of all of the following scheduled edits must also be changed.

roadblock the purchase and scheduling of a number of SPOT AN-NOUNCEMENTS on several cable or television station operations simultaneously. The objective is to obtain maximum saturation for a product or service with an advertising blitz in a small but particular time period.

robotic television cameras cameras without operators. Mounted on tracks, they can be manipulated from the control room by a single operator seated at a computer panel. CLOSE-UPS (CU) and LONG SHOTS (LS) can be preprogrammed and the camera can be directed to PAN, TILT, and ZOOM.

Rocky Mountain Network See PA-CIFIC MOUNTAIN NETWORK (PMN).

Roper Organization, The a pub-lic-opinion research firm that con-ducts national studies for individual companies to ascertain attitudes about a number of topics, including the media and communications. The firm develops public-opinion polls, researches consumer behavior, and studies marketing opportunities and strategies.

rule of thirds a television produc-tion guideline stating that the center of attention of a picture should be one-third of the way from the top of the screen or one-third of the way from the bottom, or one-third in from either edge of the screen. It should never be in the dead center of the picture.

run-of-schedule (ROS) rates the rates charged by a cable or television operation for COMMERCIAL TIME any-where within its program schedule. The advertiser or its agency buys time for the COMMERCIALS, which are then placed at the discretion of the television or cable company. That time can be anywhere from SIGN-ON to SIGN-OFF. ROS rates are lower than those for specifically scheduled time periods. (See also BEST TIME AVAILABLE [BTA], PREEMPTABLE RATES, RATE CARD, and STANDARD RATE AND DATA SERVICE.)

rundown sheet a list of the se-quence of events that is scheduled to occur during a television program. It often replaces a full script on infor-mal shows. A rundown sheet is used in TALK SHOWS or on ad lib INSTRUC-TIONAL TELEVISION (ITV) programs.

S

Satellite Broadcasting and Communications Association (SBCA) a national organization of companies involved in the TELEVISION RECEIVE ONLY (TVRO) industry. SBCA has a membership that includes retail dealers and manufacturers of home satellite receiving equipment, including dishes and DECODERS and program suppliers.

Satellite Educational Resources Consortium (SERC) a consortium consisting of PUBLIC TELEVISION (PTV) stations and state education departments from more than 20 states. It produces and delivers instruction by satellite to high school students and teachers. The DISTANCE EDUCATION project offers students the opportunity to study subjects not taught in their area schools.

satellite hit a programming term describing a program that is successful because it is scheduled next to a show with high ratings. It would fail if it left the orbit of the top-rated program.

satellite master antenna television (SMATV) system a television distribution service that is used at MULTI-DWELLING UNITS such as apartments, condominiums, hotels/motels, and hospitals. It is sometimes called "private cable." Often an extension of the MASTER ANTENNA SYSTEM (MATV) setup on the rooftop, the system brings in SATELLITE-delivered programming and distributes it to the residents. The FEDERAL COMMUNICATIONS COMMISSION (FCC) ruled that no license is required to install a TELEVISION RECEIVE ONLY (TVRO) satellite dish, and because the system is on private property and no streets are crossed to connect the individual residents, no FRANCHISE is required from the city or county.

satellite news gathering (SNG) a rapid news sending-and-receiving process. SNG became practical in the 1980s with electronic improvements and less expensive equipment. The concept was fully exploited by the CONUS COMMUNICATIONS operation of the HUBBARD BROADCASTING COMPANY in 1984. It covered news events with trucks equipped with satellite UPLINKS and sold the information to many stations with TELEVISION RECEIVE ONLY (TVRO) dishes.

As the practice spread, stations began to acquire their own uplinks, and news pools were arranged among stations in various regions. And as the necessary gear became smaller and even less expensive, many stations purchased satellite news vehicles (sometimes called SNVs or "live eyes") to transmit breaking news to their own stations or others in the region or around the nation.

satellite station a television station that operates as a full-power outlet to rebroadcast the signal from a parent station on the same or a different channel. A satellite station is usually located in a MARKET where advertising revenue cannot support a full-power station and it is not economical to originate programming from the locale.

satellites   space vehicles that orbit the earth and are capable of receiving and retransmitting voice, radio, and television signals. The most powerful are those in GEOSYNCHRONOUS ORBIT, which are launched by rockets and designed to go into orbit at exactly 22,300 miles above the equator. At this height and traveling at the same rotation speed of the earth, they are in effect stationary. Because they are always in the same spot (relative to earth), they can be effectively used to receive and transmit signals from EARTH STATIONS.

A signal is UPLINKED to the satellite and is picked up by a TRANSPONDER, which changes the FREQUENCY of the signal and DOWNLINKS it to TELEVISION RECEIVE ONLY (TVRO) dishes within the geographic receiving area (or FOOTPRINT) covered by the satellite. The TVROs are located at television stations or at the HEADENDS of cable systems and both operations can retransmit the signal to their viewing audience. Many TVROs have been purchased by individuals in rural areas beyond the reach of cable and broadcast signals. They receive their programming directly from the satellite, forming a *de facto* DIRECT BROADCAST SATELLITE (DBS) system. Most of the signals, however, are SCRAMBLED by the originators and cannot be seen without a DECODER.

Most satellites operate on one of two high microwave frequencies, the KU-BAND or the C-BAND, and are so labeled in the industry. The specific bands and orbital slots (the longitudinal degree of a satellite) are licensed by the FEDERAL COMMUNICATIONS COMMISSION (FCC) to individual satellite operating companies. (See also EARLY BIRD SATELLITE, ECHO I SATELLITE, INTERNATIONAL TELECOMMUNICATIONS UNION, and TELSTAR SATELLITE.)

saticon tube   a camera pickup tube that is part of the top-of-the-line industrial cameras used in CORPORATE TELEVISION and AUDIOVISUAL COMMUNICATIONS. It provides a better picture than the older inexpensive VIDICON TUBE.

SCA   an acronym for the Subsidiary Communication Authorization Service, an awkward FEDERAL COMMUNICATIONS COMMISSION (FCC) term for a relatively new audio channel. The Commission later changed the nomenclature to Subsidiary Communication Service (SCS), but the broadcasting industry continues to use the original term. It is the second audio, or subcarrier channel, used for stereo broadcasting and other purposes in the United States. The signal is multiplexed on the FM band by a modulator. (See also SEPARATE AUDIO PROGRAM [SAP].)

scanning line   the image on a CATHODE RAY TUBE (CRT) that is produced by the movement of a spot of light across the face of the tube. This spot of light is created by a beam of electrons from an electron gun, which generates light when the electrons strike a coating of phosphor on the inside of the face of the tube. The electron beam (and thus the spot of light) moves across the face of the tube from left to right, creating a narrow line called a scanning line. In the NTSC television standard, the electron beam draws a sequence of 262-1/2 scanning lines, from the top to the bottom on the face of the CRT, to create a FIELD. When a complete field has been drawn, the beam returns to the top of the screen and draws an additional 262-1/2 scanning lines between the first field's lines. Two successive fields, one consisting of odd-numbered scan lines and one of even-numbered lines, are required to complete a FRAME of 525 lines in the NTSC standard.

scatter market   the sale of unsold COMMERCIAL TIME in an opportunistic manner by STATION REPRESENTATIVES (REPS), FIRST-RUN SYNDICATORS, and NETWORKS throughout the year. This type of sale (usually on a quarterly basis) differs from an UPFRONT BUY in which time is sold before the television SEASON begins.

scatter plan   the practice of buying COMMERCIAL TIME on television or cable operations in a variety of programs, sometimes called a scatter buy. Advertising agencies occasionally buy such time at PREEMPTABLE RATES. Most often, the advertiser does not have any control over the times the commercials are to air and the SPOTS are often purchased at RUN-OF-SCHEDULE RATES.

scoop light   a type of floodlight used to create FILL LIGHTING over a rather large area. It is sometimes called a "basher." It is a round funnel-shaped appliance, up to two feet in diameter, with a large bulb of 1,000 to 1,500 watts designed to throw soft light over a broad area. To further diffuse the light, a spunglass SCRIM is sometimes placed over the opening to make the light even softer. Scoops are often hung in pairs on HANGERS.

scrambling   the process of encrypting an electronic signal to curb access to it. It rearranges elements of the signal into a jumble of contradictory pieces. While there is a semblance of a picture at the receiving end, the image is incoherent and not recognizable. A DESCRAMBLER is required to receive a clear picture.

Screen Actors Guild (SAG)   a trade union representing performers appearing in COMMERCIALS, motion picture films, MADE-FOR-TV MOVIES, and television programs, as well as educational and CORPORATE TELEVISION nonbroadcast programming. The union has jurisdiction over such work throughout the United States.

   SAG was formed in 1933 in Hollywood to represent actors in the film industry. It expanded its role when television was launched. (See also ALLIANCE OF MOTION PICTURE AND TELEVISION PRODUCERS and AMERICAN FEDERATION OF TELEVISION AND RADIO ARTISTS.)

Screen Extras Guild (SEG)   a trade union representing nonfeatured actors in nonspreaking roles in films and MADE-FOR-TV MOVIES or COMMERCIALS. SEG members include singers, dancers, horse riders, athletes, car drivers, and people who do stand-in work or appear in crowd scenes.

screeners preview copies of videocassette titles sent to WHOLESALERS and retail stores by the PROGRAM SUPPLIERS. The majority of distributors and retailers prefer to screen a title before they make a purchase decision.

scrim a spun-glass fibrous material that is sometimes placed in front of a lighting instrument such as a SCOOP to diffuse and soften the light. A frame containing the scrim is attached to the front of the instrument by clips.

The term is also used for a large curtain drop made of loosely woven cotton that appears opaque when it is lit from the front and translucent when lit from behind.

seamless transition See HOT SWITCH.

season the period of time in television from fall to spring. It encompasses some seven to eight months of activity in which new programs are introduced, old ones begin a new "year" of episodes, and new schedules are devised and promoted. In the early years of the medium, Labor Day was the traditional kickoff point for the new television season for the networks, but that date has changed in recent years. The new schedule now usually begins in the latter part of September.

Many shows fail and are replaced in midseason. Recognizing the high mortality rate for mediocre shows, the networks began the concept of a "second season" beginning in January. Some have begun scheduling replacements after just a few weeks, thus ushering in the second season just after the November SWEEPS. The fine line separating the fall season from the second season has therefore been eroded in the past few years. There is a growing trend toward a seasonless television year, with SYNDICATORS introducing new shows at any time and even the networks premiering shows throughout the year. (See also QUICKSILVER SCHEDULING).

SECAM the French-developed transmission and reception engineering standard used in that country, in eastern Europe, the Middle East, and the USSR. The acronym stands for "*sequential electronic couleur avec mémoire*" (sequential with electronic color memory), although some have dubbed it the "system essentially contrary to American method." The system uses 625 scanning LINES and 25 FRAMES per second except in France, which uses 819 scanning lines. It is only partly compatible with the PAL standard and totally incompatible with the NTSC standard. (See also ADVANCED TELEVISION [ATV].)

second season See SEASON.

segue   the smooth transition from one sound to another, often by the use of a CROSSFADE. Pronounced "SEG-way" but often shortened to "seg" in everyday use, this French term is used primarily in television audio production. Originally a musical direction, the term was adopted in radio to describe the easy movement without pause from one record to another. In television audio, it characterizes a production technique in which one sound fades out while another fades in. The term is less frequently used to describe visual transitions such as a change of a mood or scene, usually through a DISSOLVE.

sell-off   the practice of selling used or new videocassettes at low prices after their popularity has peaked. Retailers have always offered units that have recouped their costs but are out of current favor or "old." They are sometimes placed in bargain bins and euphemistically labeled "previously viewed."

sell-through   a technique in home video retailing in which prerecorded videocassette titles that are low-priced are planned and intended for "sell-through" (to the customer) rather than rented. The titles are often accompanied by a massive promotion, publicity, and advertising campaign to assist in their marketing at the retail level.

separate audio program (SAP)   a program service that relies on a third audio subcarrier channel. It was authorized for television stations by the FEDERAL COMMUNICATIONS COMMISSION (FCC) in 1984 as a part of the new stereo TV standards. It is offered by terrestrial television broadcast stations or via SATELLITE or cable networks. It allows for the simultaneous translation of ongoing, foreign-language programming, around-the-clock weather reports, or other audio information, separate from or in conjunction with a transmitted picture.

SESAC Inc.   a private organization representing music companies in licensing the performance of their work for use in recordings, radio, and television in the United States. SESAC is a licensor of synchronization music rights for motion pictures, syndicated television series, and commercials, as well as background music and premium record albums.

Seven Dirty Words case   a landmark legal action that was the first major attempt to address the issues of obscenity and indecency in broadcasting. It pitted the FEDERAL COMMUNICATIONS COMMISSION (FCC) against a noncommercial radio station, WBAI-FM, and its license holder, the Pacifica Foundation. In 1973 the station broadcast a recording by comedian George Carlin of his nightclub act, during which he discussed and repeated the seven words "they won't let you say on the air." He repeated the words 106 times during the 12-minute monologue.

While the language was not "obscene," it was "patently offensive," according to contemporary community standards, said the FCC, and it violated the Commission's definition of decency. The station fought the reprimand on the basis of the First Amendment, but the Supreme Court upheld the Commission.

seven-day rule   a FEDERAL COMMU-NICATIONS COMMISSION (FCC) rule that applies to the EQUAL TIME (OP-PORTUNITY) RULES relating to political broadcasting. It states that a candidate's request for an equal opportunity to broadcast a program in response to an opponent's appearance must be submitted to the station within one week.

777 rule   a FEDERAL COMMUNICA-TIONS COMMISSION (FCC) regulation limiting the number of radio and television stations that commercial companies could own. The intention was to diversify the ownership and control of the broadcast media. The FCC limited ownership by any one company to seven AM and seven FM radio stations and seven UHF and VHF TV stations. The rule was adopted in 1953–54 and was in effect until it was replaced by the 12-AND-25-PER-CENT RULE in 1984.

share   the percentage of television households in a given market that are watching a specific program or station at a particular time. A share is derived by dividing the viewing level (number of viewers) of each station or program by the total number of HOUSEHOLDS USING TELEVI-SION (HUT) in that specific time period.

As with RATINGS, shares are calculated in percentages. If there are 10,000 television sets turned on in a particular time period and 3,000 are turned to a particular program, that program would have a share of 30. Shares are always larger than ratings.

shooting ratio   the relationship between the amount of film or tape shot to the amount used in the finished project. Many in the film world believe that an ideal ratio is 10:1 (10 times as much film will be shot than will ever reach the screen after editing). Television shooting ratios are usually much lower than those for theatrical movies. The majority are in the 3:1 or even 2:1 ratio.

shot list   a rundown of the shots that are assigned to each camera. It is sometimes called a "camera cue sheet" or a "shot sheet." For complex dramatic or musical television productions, some directors number their shots on a script in sequence and give a shot list to the camera operators, who tape the narrow strip of paper to the backs of their cameras near the viewfinder. The terms "shot sheet" and "shot list" are also used to describe the brief list of shots that have been recorded on videotape. They contain time-code numbers and a short description of each shot along with any comments about it. The lists are used by editors to select shots in the VIDEOTAPE EDIT-ING process.

shotgun microphone   a highly UNI-DIRECTIONAL MICROPHONE that is often used in on-location film and remote television production. It is used to pick up sounds at a distance. The long, tubelike mike resembles a gun and is mounted on a boom or aimed at the talent by a production assistant.

shrink wrap   a process used to put a clear protective plastic covering on the box containing a blank video tape or prerecorded video title. The plastic is shrunk by heat to fit snugly around the box and protects it during shipping and handling.

shrinkage   the slow loss of INVEN-TORY over a period of time in the home video retail industry. The gradual reduction is due to damaged videocassettes, "paper errors" (such as failing to register a sale), and theft by customers or employees. Shoplifting is the major cause, followed by internal theft.

sign on/sign off   the start and conclusion of a broadcast day at a radio or television station. The FEDERAL COMMUNICATIONS COMMISSION (FCC) requires stations to broadcast STA-TION IDENTIFICATION (ID) announcements at the beginning or ending of each hour of operation. Stations sign on in the morning with a formal audio and video announcement that broadcasting is beginning on a specific channel, at a specific power, and from a specific location, as authorized by the Commission. At the end of the broadcast day, stations sign off by announcing that they are ceasing transmission.

signal-to-noise ratio (S/N)   a method of measuring the extent of the corruption of the electronic signal. BACKGROUND NOISE or distortion is an inherent part of any electronic system. By using an algebraic formula, $S/N(dB) = 10 \log_{10} (P_m/P_n)$, engineers can compare the strength of the signal emanating from an electronic apparatus with the equipment's internal disruptive electronic forces. The relationship is expressed in DECIBELS (DB). The higher the S/N ratio, the cleaner the signal. A signal-to-noise ratio of 0 dB would mean that the background noise in that device has the same power as the signal and the result is an extremely distorted signal.

simulcast   the **simul**taneous broad**cast** of the same program over two different channels. Many AM radio stations simulcast their programming on sister FM stations. Some co-owned public radio and television stations simulcast an opera or symphonic concert on AM radio and also use the FM radio station's regular stereo channel, thereby providing two radio and one television audio channels for the stereo sound accompanying the television picture.

simulcast high definition television (HDTV)   types of proposed AD-VANCED TELEVISION (ATV) transmission systems that are not based on the current NTSC television standard, but are designed to operate in a parallel pattern with it. They are therefore called SIMULCAST systems. Because they increase the number of SCANNING LINES and FRAMES per second, they are also called "high definition television (HDTV)" systems.

The proposed systems will use the same six MHZ wide-channel system with the signals transmitted simultaneously with current NTSC signals. Thus the same program will be transmitted at the same time in NTSC and in HDTV. The separate HDTV signals, however, will require a different television set to receive that signal. Eventually the separate NTSC transmissions would not be needed, as viewers replace their existing TV sets with HDTV sets.

sitcoms light television dramas focusing on the humorous side of supposedly real-life circumstances, usually centered around family life. The term is short for situation comedy. The 30-minute weekly story is told in three acts, with the first eight minutes setting up the situation, the second eight minutes complicating it, and the final eight minutes devoted to the comedic resolution. (The other six minutes are devoted to COMMERCIALS.) The plots concern themselves with everyday occurrences involving sympathetic people who have distinguishing characteristics.

Some cater to a particular audience segment: male ("Major Dad"), female ("Designing Women"), or older ("Golden Girls"). While a popular, charismatic star is important, distinctive supporting characters and quality writing ensure long-term popularity and durability. Most connoisseurs cite "I Love Lucy" and "All in the Family" as the best of the lot in

terms of broad humor, and "Dick Van Dyke," "Mary Tyler Moore," "M*A*S*H," and "Cheers," best in empathy.

Almost every country has its versions of situation comedy. A few are exported, notably those from the United Kingdom, which are sometimes called "Britcoms," occasionally "witcoms," and less reverently "twitcoms" when they are televised on PUBLIC TELEVISION (PTV) stations in the United States.

Sixth Report and Order a FEDERAL COMMUNICATIONS COMMISSION (FCC) report that was the basis for television service in the United States. Issued on April 14, 1952, the document established a table of allocations for channels throughout the nation and lifted THE FREEZE on the construction of new stations. The report was the final one (of six) in a comprehensive technical and economic study by the Commission that examined all of the various implications in channel allocations. Altogether, 2,053 channel allocations were made to 1,291 communities throughout the United States. The FCC reserved 242 channels for the specific use of noncommercial EDUCATIONAL TELEVISION (ETV). Most were in the UHF band, joining the 66 percent of all the allotments that were in that new service.

In spite of the intermixture difficulties and other problems, the Order became the charter for the growth of the U.S. television system. Only 120 stations were on the air in 1952 but by the next year the number had doubled, and by 1955 some

400 stations were operating. Today more than 1,300 stations serve the American public.

Skycam a production device in which a camera is suspended on wires over an arena or stadium. The wires are connected to small winches, which are computer-controlled. By adjusting the wires, the unmanned camera can be made to move and zoom from ground level to 150 feet at a maximum speed of 20 miles per hour. (See also STEADICAM.)

slant-track videotape format See HELICAL-SCAN VIDEOTAPE RECORDING.

slobcoms SITCOMS that focus on the goings-on of bickering blue-collar families. The word was coined by *Broadcasting* magazine in 1988. "Roseanne" and "Married . . . with Children" are examples of the GENRE.

smearing effect a technical aberration and disturbance in a television picture, which often occurs when inexpensive television cameras or CAMCORDERS are used for home videotape recording or the camera is used in high-contrast lighting situations. The undesired effect consists of a blurring of the vertical edges of an object. The effect is similar to the condition called STREAKING, where the same effect occurs horizontally.

SMOG grading a sophisticated index of text readability. Sometimes used by advertising copywriters, the acronym stands for "simple measure of gobbledygook." The SMOG index samples 10 sentences from the beginning, middle, and end of a prose passage, counts the number of words of three or more syllables, and determines the square root of that number to come up with the SMOG grading. The lower the score, the more "readable" the text.

SNG See SATELLITE NEWS GATHERING (SNG).

soap operas (soaps) a television series that consists of ongoing episodes with continuous storylines rather than individual, self-contained programs. The daily daytime dramas on early radio were given the name by the TRADE MAGAZINE *Variety,* because they were, in the main, sponsored by many of the leading detergents of the day. The shows are targeted at housewives.

Most daytime television soaps are one-half hour in length. The stories frequently involve such subjects as illegitimate children, seductions, rape, incest, innocent virgins, and amnesia. There is a great deal of villainous behavior and many series feature dual roles involving identical twins and split personalities. Some critics maintain that the soaps are a valid form of theater and as legitimate as Shakespeare's plays, while others dismiss them as drivel.

Latin American versions of soap operas called *TELENOVELAS* are even more popular than those produced in the United States, but the episodic dramas are produced throughout the world. In French Canada the dramas are known as "teleromans" or "TV novels." The daily doses of strife in the United Kingdom are usually called "serials."

Society for Collegiate Journalists (SCJ) a nonprofit organization of students in print and broadcast journalism. It has many chapters at colleges and universities and funds research programs and contests and evaluates higher education programs in journalism.

Society of Broadcast Engineers (SBE) a nonprofit professional society serving the interests of broadcast engineers. It conducts seminars on electronics and technical topics and operates a certification program.

Society of Cable Television Engineers (SCTE) a professional membership organization consisting of cable television engineers and technicians. The group provides a forum for information exchange, sponsors workshops, and encourages young people to enter the field.

Society of Motion Picture and Television Engineers (SMPTE) the oldest and most prestigious organization of professional engineers and technicians. Founded in 1916, the society works to further technology, distributes scientific information, and encourages the practice of engineering. The group has been particularly active in developing SMPTE technical standards for motion pictures and television. (See also AD-VANCED TELEVISION [ATV].)

Society of Professional Journalists (Sigma Delta Chi) a professional society that holds seminars in print, radio, and television news and sponsors internships and symposia on the press. The society publishes a monthly magazine *(The Quill)* and holds regional professional development conferences each year.

Society of Satellite Professionals International (SSPI) a nonprofit organization that seeks to create a worldwide network of personal contacts to promote professionalism in the satellite industry and to increase awareness of the field. The membership consists of individuals that have college degrees and three years of experience in the field. Individuals without a degree must have seven years of experience.

soft light subtle, smooth, gentle light for a television or film production. It is created by large SCOOPS that are heavily diffused or by bouncing light off reflectors.

software prerecorded video titles that are duplicated and sold to consumers. Videotapes are software, as opposed to equipment such as videocassette machines, which are considered HARDWARE.

Sony Corporation one of the world's leading manufacturers of personal audio and video equipment, television sets, semiconductors, and computers. The company is a worldwide giant in electronics with 576 consolidated subsidiaries and three affiliated companies. The firm pioneered in audio consumer products and in the development of videotape recording and reel-to-reel VIDEOTAPE FORMATS. In 1975 the firm introduced the first videocassette machine, which was labeled the BETA FORMAT. Although it lost the format war to the VHS FORMAT developed by its rival MATSUSHITA, the firm prospered in other fields.

Among the consumer products developed and sold under the Sony name have been the audio and video Walkman, TV projectors and sets, the Handycam 8mm video recorder, audio COMPACT DISC (CD), and LASER VIDEODISC (LV) players. The company is also known for its professional broadcast equipment including BETACAM cameras and other studio gear. The company has also introduced CHARGE-COUPLED DEVICE (CCD) cameras and ADVANCED TELEVISION (ATV) monitors and projectors.

In 1988 it bought CBS Records Inc. to enhance its audio product line and in 1989 the company purchased COLUMBIA PICTURES ENTERTAINMENT (CPE) (renaming it Sony Pictures Entertainment), to become the first Japanese owner of a large Hollywood entertainment company. In 1991 the company bought out General Electric's interest in RCA/Columbia Pictures Home Video.

sound bites 10- or 15-second snippets of speech used in television news reports. They encapsulate a mood, an issue, or a politician's attitudes and beliefs and are the blurbs, quips, and slogans characteristic of modern communication. The term became popular with the practice of editing television interviews based on what an individual says rather than the visual image. Because an image of someone talking is less dynamic, only their words are considered newsworthy and short excerpts can be extracted from long speeches. Consequently political candidates articulate their beliefs in short, simple, memorable, and clearly understood phrases. The pithy one or two sentences are the right length to hold the attention of a modern audience. Reporters and news directors thrive on sound bites, and the punchy character they embody enlivens a report and helps maintain a fast-paced newscast.

soundies big-band-era film shorts that were the precursors of today's music videos. Introduced into the home video market in 1990, they are each approximately three minutes in length. The films were once run on "visual jukeboxes" in the late 1930s and 1940s and are now available on videocassette.

Southern Educational Communications Association (SECA) a nonprofit regional organization serving both PUBLIC TELEVISION and public radio stations in the southern part of the United States. Through its acquisitions and distribution, SECA provides its member stations with regional and national programming, often through a GROUP BUY. SECA also provides services in INSTRUCTIONAL TELEVISION programming and professional development.

space coverage television coverage of space exploration and activities. The networks began their coverage of manned space missions as early as 1960, but it was the first orbital trip in February 1962 that completely captured the imagination of the American public. In May 1963, a view of the earth ("the big, blue marble") from space captivated viewers as television broadcast the first space walk. In 1967 (on Christmas Eve) an astronaut read the Bible to millions of viewers as a spacecraft orbited the moon.

On July 20, 1969, television broadcast the incredible sight of men walking on the moon. It was seen by 723 million people in 47 countries, the largest television audience for any single telecast. The Challenger space shuttle flight in 1986, however, was a tragedy. The spacecraft exploded 70 seconds after takeoff, killing all aboard. The disaster unfolded live before millions of horrified people. The accident forced a halt in the shuttle launches until 1988 and few have been televised since then.

Spacenet satellites See GTE SPACENET CORPORATION.

Spanish International Network (SIN) See UNIVISION.

special-effects generator (SEG) See DIGITAL VIDEO EFFECTS and SWITCHER.

special event video (SEV) the coverage of weddings, recitals, graduations, and pageants by small video production companies. Using PROSUMER equipment and DESKTOP VIDEO techniques, the one- and two-person firms compile video yearbooks, cover funerals, record depositions, and record some of the more than 4,000 beauty pageants held each year in the United States.

special interest (SI) programming programs that seek a particular TARGET AUDIENCE for a particular subject. The topics covered are diverse and the FORMATS vary from DOCUMENTARY to NEWS-TALK and lecture-demonstration. Self-improvement programs or HOW-TOS are a basic part of special interest programming in subject areas ranging from auto repair to aerobics. Foreign language lessons and programs about boating, child care, fashion, gardening, and money management are all special interest shows, as are travelogs and programs on specific hobbies or crafts. The SPECIAL INTEREST VIDEO ASSOCIATION (SIVA) broadly defines the term as everything on video except theatrical and motion picture features. The term is often used synonymously with NONTHEATRICAL FILMS/PROGRAMS/VIDEOS.

Special Interest Video Association (SIVA) a nonprofit association composed of individuals and companies involved in the production and distribution of SPECIAL INTEREST (SI) PROGRAMMING. SIVA specifically defines special interest videos as those that "teach, train, guide, direct, inform, inspire, entertain, enlighten, and enrich those who use the programs."

special temporary authorization (STA) the permission granted by the FEDERAL COMMUNICATIONS COM-MISSION (FCC) to a station to broadcast at a specific variance from the conditions and terms of its FCC LICENSE. An STA may also be granted by the FCC when a radio or television station requests permission to temporarily deviate from Commission requirements for a particular class of station. In either case the STA is given for a limited time ranging from three to six months, depending on the circumstances.

specials single programs that are usually scheduled in PRIME TIME on an occasional basis to attract a large audience. Sometimes called a one-time-only (OTO) program, the hour-long, 90-minute, two-hour, or longer show is usually an entertainment *tour de force* and is heavily promoted. The show can be a yearly event or a reunion program featuring performers and FILM CLIPS from previous series. Because there are so many of them on both cable and broadcast television today, the term "special" has lost its initial appeal. (See also SPECTACULARS.)

spectaculars elaborate and expensive SPECIALS that brought new viewers to the small screen in the early days of television. Although programming executives were hesitant to break into the viewing routines of the regular audience, they occasionally tried a one-time-only show to attract a new audience. The single shows were usually 90 minutes in length and were extravaganzas featuring brilliant costumes, lights, sets, and production numbers. The new

talent and stars attracted huge audiences. As their number grew, however, the shows became less and less spectacular and unusual, and the term gradually gave way to the more humble "special," which in turn has become commonplace and lost much of its appeal.

Speech Communication Association (SCA) a nonprofit association, headquartered in the Washington D.C. area, that is dedicated to the communication arts and sciences in higher education. Its members include faculty and students from colleges and universities as well as the institutions themselves. Members in its Mass Communications division are, in the main, college teachers of radio and television who also do research in the field.

spin-off a program that usually features the secondary stars of a hit SITCOM in new circumstances or surroundings. Many of commercial television's best-known programs were developed as spin-offs. "The Mary Tyler Moore Show" spun off both "Rhoda" and the less successful "Phyllis," while "All in the Family" spun off "Maude," which spun off "Good Times." It is almost axiomatic in commercial television today that a successful show with strong characters will create at least one spin-off.

split screen See WIPE.

split-barter syndication a type of BARTER SYNDICATION in which the television or cable operation and the program distributor divide the COM-

MERCIAL TIME available in a program. The practice is sometimes called "barter-split." Today it is used extensively for FIRST-RUN SYNDICATED PROGRAMS both in cable and broadcast television. Typically, the time available is split 50-50.

split-format programming a type of programming in which two different but related stories are shown in one hour. The two-shows-in-one program can be used in SYNDICATION as an hour show or as two half-hours of programming. One variation has the drama seen from two different perspectives, such as that of a husband and a wife. Another has a complete story of the first 12 hours of a day in the first half-hour, followed by the second 12 hours in the second half-hour.

sponsor a company that pays for all of the COMMERCIAL TIME during a television or cable program. At one time the sponsor also financed the production, but this occurs only on occasion today. The company seeks the prestige of close identification with a quality program or a star. In this way, they control the program content as well as the placement of COMMERCIALS within the program. Sponsorship is usually reserved for an event or SPECIAL such as the dramas on "Hallmark Hall of Fame." (See also INFOMERCIALS.)

sponsored video a video title that is tied to a related nonvideo product or service for advertising purposes. Most are SPECIAL INTEREST (SI) programs.

The simplest form of sponsorship is for companies to offer the title as a premium to encourage customers to buy their product. This type of participation usually does not engage the advertiser directly in the production or marketing of the title. Increasingly common, however, are the arrangements by which an advertiser pays for some of the production costs in return for a BILLBOARD at the beginning or end of the program. Some sponsored titles feature the same 30- or 60-second SPOTS that are made for commercial television.

sponsorship identification rules Section 317 of the COMMUNICATIONS ACT OF 1934, which requires that a broadcast station that transmits a paid-for program must announce the fact that the program is paid for or sponsored. The radio or TV station must also announce who paid for the program. If a program is on a political or controversial issue and was provided to the station without charge, the station must announce that fact, as well as the name of the organization or individual that provided the program. (See also UNDERWRITING RULES.)

sports blackout rules FEDERAL COMMUNICATIONS COMMISSION (FCC) rules that limit the telecasting of a sporting event to the locale where the event takes place. The event may not be broadcast locally if it is not sold out. If the event is not being carried on a broadcast television station in that market, it cannot be carried by a cable system. The rules are designed to protect those owning the television rights to games, because if the event is televised and seen locally, there is often a decrease in ticket sales.

SportsChannel America   a BASIC CABLE network that concentrates on prerecorded and live sports programming. It is owned by NBC and CABLEVISION SYSTEMS CORPORATION.

spot (commercial time)   the time between or within programs on an individual television station or cable system during which COMMERCIALS are shown. The commercial time is sold by the station or system to local ADVERTISING AGENCIES or firms. Spot time is also sold by STATION REPRESENTATIVES (REPS) to national advertisers through national ad agencies that purchase the time on an individual station or cable system in more than one market. (See also MARKET-BY-MARKET BUY, SPOT [CONTRACT], and UNWIRED NETWORKS.)

spot (announcement)   a short COMMERCIAL that is placed on a cable or television operation between or within programs. Spots are usually less than one minute in length and commonly have a 10-, 15-, 30-, 45- or 60-second format.

spot (contract)   an advertising contract for the purchase of COMMERCIAL TIME in a local television MARKET or on a local cable system. The contract details the nature and terms of the purchase including its cost, the number of SPOT ANNOUNCEMENTS to be run over a period of time (such as a week or a month), and the times the

spots will air. (See also MARKET-BY-MARKET BUY and UNWIRED NETWORKS.)

spotlights   lighting devices used in television or film productions. Informally called "spots," these instruments are used for KEY LIGHTING and BACKLIGHTING, both in the studio and on REMOTE. They provide the brightest and most concentrated light and create the most shadows. Spotlights are employed to cover a very limited area. They are usually rectangular instruments with a round glass lens and often have BARN DOORS attached to them to further concentrate their light.

stand-alone   an early cable television term that described cable systems that were not part of any cable NETWORK. While the systems carried local and DISTANT SIGNALS, they secured videotapes directly from distributors and other program suppliers, often via a BICYCLING method, and ran their own individual PAY (PREMIUM) CABLE SERVICES. Lower-cost TELEVISION RECEIVE ONLY (TVRO) dishes made interconnection via satellite feasible, and by the early 1980s nearly all stand-alone operations had been replaced by the programming services from satellite cable networks.

Standard Rate and Data Service (SRDS)   a publisher of the RATES of television stations throughout the United States, known informally as SRDS. The information, published monthly, is used by companies and their ADVERTISING AGENCIES to price

SPOTS for possible purchase. (See also RED BOOKS.)

standups brief televised news reports by reporters in the field. The on-air reporters position themselves standing by a burning building, at the scene of an accident, or in front of the Capitol or White House and, speaking to the camera, report directly to the folks at home.

star network the basic wired interconnection system that is the standard in the telephone industry. In the star configuration, wires from a central point are connected to each individual telephone subscriber's home. Each customer has a private line. This design contrasts with the shared-line technique in the TREE NETWORK method that is common in the cable industry. Star network configurations, however, will probably be used when FIBER OPTICS technology is totally adopted by the cable industry. A nearly unlimited number of channels can be sent from a headend to each individual home, one at a time, at the subscriber's request.

star schools program an educational project that seeks to use SATELLITE and other technology to equalize learning opportunities for the nation's youth. Administered by the U.S. Department of Education, the program awards two-year grants to regional projects to develop DISTANCE EDUCATION programming and services. The projects are designed to reach rural schools that do not have advanced (or sometimes any) instruction in some subject areas. Most of the projects involve other technology (including LASER VIDEODISCS [LV) and use COMPUTER-ASSISTED INSTRUCTION (CAI) and INTERACTIVE TELEVISION techniques to link teachers and students. (See also SATELLITE EDUCATIONAL RESOURCES CONSORTIUM and TECHNICAL EDUCATIONAL RESEARCH CENTER.)

state networks noncommercial PUBLIC TELEVISION (PTV) networks or commercial networks that cover a state. The noncommercial stations are licensed by the FEDERAL COMMUNICATIONS COMMISSION (FCC) to a state authority or agency and one main station feeds a network of SATELLITE stations. A network may also be formed within a state by commercial stations that are owned and operated by a single company and are interconnected by MICROWAVE RELAY.

station breaks the period between programs in a broadcast operation. Most stations use the time to comply with the FEDERAL COMMUNICATIONS COMMISSION (FCC) requirement that all radio and television broadcast stations publicly identify themselves periodically. The STATION IDENTIFICATIONS (ID) consist of the transmission of the CALL LETTERS, CHANNEL, and the city of origin, both visually and aurally. The station break periods are also used as COMMERCIAL TIME slots. (See also CONTINUITY and PROMOS.)

station identification (ID) an announcement on a broadcast station consisting of the station's CALL LET-

TERS, CHANNEL, and city of origin. Stations are required by the FEDERAL COMMUNICATIONS COMMISSION (FCC) to periodically publicly identify themselves. The station IDs must be aired at hourly intervals or, if a program runs longer, at a natural break near the hour. IDs are designed by the FCC to minimize any potential confusion among viewers about which station they are tuned to. (See also SIGN ON/SIGN OFF.)

station lineup the number of television stations that have committed to license and carry a syndicated program. The *ad hoc* noninterconnected group of stations is usually created for FIRST-RUN SYNDICATED PROGRAMS, but OFF-NETWORK shows are also sold to a lineup of both AFFILIATED and INDEPENDENT STATIONS. No two programs will have exactly the same station lineup.

station representatives (reps) national companies that sell COMMERCIAL TIME on local stations to national advertisers. The rep firms are usually located in New York City but have offices in many large MARKETS, where major ADVERTISING AGENCIES and their CLIENTS are located. They represent a number of stations in selling SPOTS from AVAILABILITIES, thereby offering national advertisers a MARKET-BY-MARKET BUY. Some organize UNWIRED NETWORKS. (See also SPOT [CONTRACT], SCATTER PLAN, and STATION REPRESENTATIVES ASSOCIATION INC.)

Station Representatives Association Inc. a national association of STATION REPRESENTATIVES (REPS) from the national firms engaged in representing local commercial television stations in SPOT television sales. The association was founded in September 1947 and is headquartered in New York City.

Statistics Canada a government agency that compiles, analyzes, abstracts, and publishes statistical information on the economic and social life of Canada and conducts a nationwide census every five years. It is similar to the Census Bureau in the United States.

Steadicam a device consisting of a small platform mount capable of holding a camera. It has a support arm that attaches to the operator's vest. The arm can be moved in any direction, powered by a battery in a module below the mount. The mount is controlled by a unit on the operator's waist, and a system of counterweights allows the camera to be moved smoothly and steadily, eliminating the slightest jerk from the picture. The remarkable assemblage makes it possible to achieve the steadiness and soaring effects usually obtained only by CRANE and DOLLY shots. (See also SKYCAM.)

step-up fees additional charges that are often required when a program has been produced for and aired on a local station and is later acquired by a network for national RELEASE. Most step-up fees are necessary because of the contractual arrangements related to RESIDUALS. Although original fees are paid to writers, actors, directors, and other creative people as well as technical personnel for the initial distribution of the program or film, other charges are incurred when the program "steps up" to wider distribution.

still video camera   a camera designed to store still pictures on small magnetic floppy discs, which can later be displayed on a television set. The disc is approximately the size of a 35 mm slide and is known as a "video floppy (VF)." The device is capable of producing still picture images on the TV screen that exceed normal VHS standards and approach photographic quality. (See also KODAK STILL-PICTURE PROCESS.)

still video printer   a machine that makes a photographic print from a single black-and-white or color video image taken from a still video floppy disc. Prints are available in three-by-four-inch or other standard photographic sizes. Designed to accompany STILL VIDEO CAMERAS, these devices provide a way in which the images from a still video camera that are normally seen only on a television set can also be turned into a physical print. (See also KODAK STILL PICTURE PROCESS.)

stock balancing   a retail practice that records the activity of all of the titles in INVENTORY at a video retail store. The records are examined at periodic intervals and the merchant can exchange less popular titles for others in the WHOLESALER'S catalog. (See also BREADTH AND DEPTH and 20/80 RULE.)

stock shots   still photographs or FILM CLIPS of common scenes, such as waves crashing on a beach or an airplane taking off. They are obtained from stock library companies that house thousands of such scenes.

stop action   a still or freeze-frame technique that was adapted from motion picture film production to the television medium. It is used in televised football games, in which a playback of the action is stopped at a given point and the position of the players noted and analyzed. Although the technique had long been used by coaches, television took it out of the locker room and made it public. (See also ISOLATED CAMERA.)

storefront   a retail outlet that fronts on the street. The term is used to distinguish such operations from those in malls or other enclosed areas. The store-front type of retail outlet has proven to be the most successful in the home video industry.

storyboards   drawings of the visual aspects of the major scenes in a television or film production. The series of sequential sketches indicate the elements of action in the piece, based on the script, and help visualize what the final production will look like. Storyboards are commonly used in Hollywood film productions, but in television they are most often developed for commercials by ADVERTISING AGENCIES, who use them to get CLIENT approval before beginning to a shoot a SPOT. (See also BOARD IT UP.)

streaking effect   a technical aberration in a television image in which objects within a picture appear to be blurred. It often occurs in high-contrast situations such as the panning of a camera from a lighted portion of the set to a dark backstage area. In streaking, the subject appears to stretch horizontally rather than vertically, as is the case in the similar condition called SMEARING.

street date   the day on which a pre-recorded video title may be officially sold or rented by a retailer. The date is set by the PROGRAM SUPPLIER and WHOLESALERS to try to ensure that the units are shipped and arrive at the retailers' stores on that date or that they are available to them on a "will-call" basis at the wholesaler's warehouse. Most program suppliers establish a certain weekday as their street date to provide equal access to new titles for all retailers and to maximize promotion and publicity.

stripping   a program scheduling strategy in which individual programs from a series are transmitted at the same time each weekday. The programs are said to be stripped, or run ACROSS THE BOARD.

studio pedestal   a camera mount that consists of a large heavy pedestal that contains a system of counterweights, springs, and pulleys that easily allows the mounted camera to be raised and lowered manually or by pneumatic pressure. The base contains a dolly with large enclosed wheels, which are steered synchronously by a large wheel surrounding

the pedestal. The entire device is so sturdy that it can be used to achieve smooth DOLLY IN/DOLLY OUT, TRUCK, and even ARC shots.

studio-transmitter link (STL)   a specific application of a MICROWAVE RELAY system that connects the studio of a television station to its TRANS-MITTER site. The STL transmits a signal point-to-point from the studio to the transmitter where it is retransmitted on the station's assigned channel. In cable television a similar electronic configuration is called a "studio-HEADEND link (SHL). (See also ANTENNA HEIGHT ABOVE AVER-AGE TERRAIN.)

stunting   a program scheduling technique in which unusual programming is devised and transmitted, and schedule changes and preemptions are made, in an effort to draw attention away from competing channels. It is often practiced by local stations and networks, particularly at the beginning of the fall SEASON or during SWEEP periods. It is periodically condemned by many in the industry who believe that RATINGS are artificially skewed by the practice. (See also QUICKSILVER SCHEDULING.)

subliminal advertising   a now-illegal form of television advertising in which words containing an advertising message were flashed on the screen so rapidly that the viewer was not aware of them. The attempt was to capitalize on the phenomenon of subliminal perception, which allows the mind to absorb messages unconsciously. A few advertisers experimented with it in the 1950s, but it is now forbidden by FEDERAL COMMUNI-CATIONS COMMISSION (FCC) rules.

subscription television (STV) a now-defunct form of PAY TV in which the signal of a conventional full-power television broadcast station was SCRAMBLED. Only those viewers who purchased or leased a DESCRAM-BLER/DECODER could receive the transmitted program. The station's income came from subscriber fees. Because the signals were provided over the air on otherwise regular television stations, the operations were licensed and controlled by the FED-ERAL COMMUNICATIONS COMMISSION (FCC).

An STV station usually functioned as an INDEPENDENT STATION during the daytime (supported by COMMER-CIALS) and switched to an STV service during the evening hours. The first nonexperimental STV station went on the air in 1977. By the mid-1980s, however, most operations had been terminated with the stations reverting to full-time, regular, independent, commercial operations. By 1986 no STV stations remained in operation. (See also SUBSCRIPTION TELEVISION INC. EXPER-IMENT.)

Subscription Television Inc. experiment an early experiment in PAY TV that began in 1962 in the Los Angeles area. The system offered free television over a commercial station, with cable subscribers who paid a fee receiving three additional channels. It thus became an early PAY (PREMIUM) CABLE SERVICE operation. The system offered many baseball

games, motion pictures, and some educational and cultural program. In November 1964, voters passed a referendum by nearly a two-to-one vote prohibiting pay-TV in California. Although the state supreme court later ruled the referendum unconstitutional, the experiment had lost millions by that time and was bankrupt and out of business.

Sunday afternoon ghetto the period when many cultural and public affairs programs were scheduled on the commercial networks and on individual stations during the 1950s. Local interview and panel shows that did not attract SPONSORS were relegated to the then-low viewing period of Sunday afternoons. The SUS-TAINING programs had a loyal but small audience.

superimposition a production technique in which one image is imposed on (or placed over) another image. Like a double exposure in photography the resulting combination is a completely different single image. In television a "super" is accomplished by placing both levers of a SWITCHER in a half-mast position, thereby getting one-half of each camera's signal. The technique is really a DISSOLVE halted at the midway point and is often used to emphasize particular objects. A sharper, cleaner effect is achieved by KEYING an image (such as lettering) over the main image and this technique is used more often today in television production.

superstations television stations that operate locally but are also carried by SATELLITE and picked up and retransmitted by cable systems. They are called superstations because they reach millions of viewers throughout the United States. The INDEPENDENT STATIONS offer special sports programming, feature films, and syndicated programs and they effectively COUNTERPROGRAM both AFFILIATED STATIONS and other cable networks. There are seven stations in the United States that claim the status of superstation.

sustaining programs programs that are broadcast without advertising support on the commercial TV networks or local stations or on cable systems. The media operations sustain the costs of production and air time. Most sustaining programs are seen in the DAYTIME hours. They are religious or public affairs shows or news programs and some, such as presidential news conferences, are televised live.

sweeps four month-long rating periods during the year when A. C. NIELSEN gathers audience data from nearly all local television markets in the nation. The number of times the viewers of local television stations and network programs, as well as cable audiences, are measured normally varies, with larger markets surveyed more frequently. The largest are now measured continuously. All markets, however, are "swept" in February, May, July, and November of each year. (See also HYPING.)

sweetening the manipulation of an audio track to improve it. Sound effects, different voices, and music are added to the mixture in a variety of ways including OVERDUBBING. Frequencies are also often altered or boosted to touch up and enhance the audio. The term is also occasionally used to describe the improvement of the video image in a production.

switcher an electronic unit that receives signals from various sources, mixes and sorts them out, and then transmits them. There are three basic kinds of switchers used in television: production, on-air, and routing. A "studio production switcher—called a "vision mixer" in the United Kingdom—allows the director to choose from various input sources to create a television program. The unit is usually located in a studio control room and is connected to cameras, videotape recorders, FILM CHAINS, and other image sources. It consists of rows of buttons and a fader lever. More expensive production switchers contain SPECIAL-EFFECTS GENERATORS (SEG), which can create vertical or horizontal WIPES, matte and chromo KEYS, inserts, and SPLIT-SCREEN effects.

An on-air or master control switcher selects one finished program from such sources as a videotape machine, the studio production switcher (with the output of a live show) or a satellite feed, and sends the chosen signal to the TRANSMITTER for broadcast. This switcher also contains the sound of the finished program.

The third type of switcher, called a routing switcher, is similar to the master control switcher in that it transfers audio and video signals or finished programs from various sources to different locations within a cable system or a CLOSED-CIRCUIT TELEVISION (CCTV) operation.

sync an abbreviation for "synchronization." It is the video signal that directs and controls the rate of repetition of the electron gun in a camera pickup tube in its scanning pattern. It directs the gun to return to a new FIELD and holds the image steady.

Synchronization signals are also used in CATHODE RAY TUBE receivers in the form of timing pulses, which lock in the electron beam both horizontally and vertically, so that the pictures will not roll or be plagued with SKEWING or JITTER.

Sync signals also conform the signals of other electronic gear so that all of them work together in the same timeframe. Cameras, the switcher, videotape recorders, and other video system components are often connected by a SYNCHRONIZATION (SYNC) GENERATOR that maintains the operation of each unit in coordination with each of the other units.

synchronization (snyc) generator an electronic apparatus that stabilizes the television signal by means of synchronization (SYNC). It establishes the basic signal that coordinates the timing of all of the other signals in a television system. In a television production studio, the unit synchronizes the signals coming from different sources, such as two or three cameras.

Sync generators are also used to coordinate signals from a variety of sources such as cameras, FILM CHAINS, and videotape recorders. In transmitting video and audio signals through the air, the sync generator creates a third electronic signal called "sync" that keeps the video signal stable. (See also COMPOSITE VIDEO SYSTEM/RECORDING, MODULATION, and RF.)

syndex See SYNDICATION EXCLUSIVITY RULES.

syndication the practice of selling the license to air television programs on cable and television operations. The FEDERAL COMMUNICATIONS COMMISSION (FCC) defines a syndicated program as "any program sold, licensed, or distributed or offered to television station licensees in more than one market within the United States for noninterconnected (non-network) television broadcast exhibitions, but not including live presentations." The stations or cable systems license the rights to transmit programs a specific number of times over a stipulated period. After the duration of the contract, the license rights revert back to the syndicator or producer. (See also ACROSS THE BOARD, BICYCLE, CASH SYNDICATION, COMMERCIAL TIME, FIRST RUN, FULL BARTER, NIELSEN STATION INDEX, OFF-NETWORK, PRIME TIME ACCESS RULE, SPOT, STATION LINEUP, SYNDICATION EXCLUSIVITY RULES, and TIME BANKING.)

syndication exclusivity rules rules affording some programming protection for television broadcasters in an increasingly cabled America. They are commonly known as "Syndex." In conjunction with the NONDUPLICA-

TION RULES, they were reinstated by the FEDERAL COMMUNICATIONS COMMISSION (FCC) effective January 1, 1990.

The rules require cable operators to delete from their carriage duplicative syndicated programs in the local station's market. The same program imported from a distant station, cable network, or SUPERSTATION must be blacked out if the local station has a contract with the distributor of the syndicated program for exclusive rights for the program within the "geographic zone" of the broadcaster or distributor.

syndicator   an operation that sells the license for and delivers television programs to a station, cable operation, or nonprofit organization. The company or individual, often called a "distributor," acts as an intermediary sales agent between the producers of the film or program (and their agents) and the outlet that will show the program to the public. The distributor circulates the programs on videotape, film, or by SATELLITE or BICYCLING. The television syndicator differs to some degree from a WHOLESALER or distributor in the home video industry. The actual ownership of the physical product seldom changes hands in television syndication as it does in video. (See also OFF-NETWORK and FIRST-RUN SYNDICATED PROGRAMS.)

T

tabloid TV programming   a form of TALK-SHOW PROGRAMMING, so named because it resembles tabloid newspapers in the exploitation of sensational events, personalities, or topics. The shows cover provocative subjects and aggressively schedule controversial guests who discuss once-taboo topics.

A recent TV phenomenon, the tabloid programs have taken advantage of the public's increasingly liberal attitudes toward sex. More modest versions of the GENRE use a MAGAZINE FORMAT and more restrained hosts. Such shows are often touted as REALITY PROGRAMMING but their topics are often similar to the more sensational tabloid talk shows. They are normally only available in SYNDICATION.

talk-show programming   a basic television format involving a host and a sidekick/foil, a small band, and a number of guests who either perform or are interviewed in front of a studio audience. The stage setting is a desk and couch or the reproduction of an informal living room, and viewers are encouraged to feel as if they are eavesdropping on conversations in the host's own personal domain.

target audience   the television or cable viewers that a producer or advertiser seeks to reach with a program or COMMERCIAL. ADVERTISING AGENCIES use DEMOGRAPHIC and PSYCHOGRAPHIC studies containing audience profiles and CLUSTER ANALYSIS in order to purchase COMMERCIAL TIME to reach specific target audiences that they hope will buy their CLIENTS' products. Target audiences may be identified by such characteristics as attitudes, politics, and life styles.

teaser intriguing scenes or a montage from a forthcoming show, sometimes shown before the opening credits of a television program to arouse viewer interest and curiosity. The dynamic look of the BIT or action encourages the audience to stay tuned to see more.

Technical Education Research Center (TERC) a cooperative effort of 11 educational institutions to provide high school students with instruction in science and mathematics. The DISTANCE EDUCATION project reaches students in every state.

techthusiasts technically adventuresome people who strive to be on the cutting edge of consumer electronics products. They are usually affluent—with cellular phones, computers, CD-ROM players, camcorders, and any new device. The high-tech mavens are comfortable with technology and strive to be the first on the block to possess a new gizmo. (See also VIDEOPHILE.)

telecast a term sometimes applied to the transmission of a television program. It was coined in the mid-1950s by network promotion and publicity people and newspaper reporters. Replacing the prefix "broad" with "tele" effectively separates the new medium from the older radio term while still retaining the action connotation of "cast." When made-for-cable programming was introduced in the 1970s, writers and publicists began using the term "cablecasting" for shows initially released in that medium.

telecine See FILM CHAIN.

Telecommunications Research and Action Center (TRAC) a non-profit citizens' group dedicated to the improvement of radio, television, cable, telephone, and new electronic media. It represents other organizations before federal agencies (such as the FEDERAL COMMUNICATIONS COMMISSION [FCC]) and Congress.

teleconferencing an umbrella term describing a communications technique that uses technology to bring people from many locations together to share information and an interchange of views. The communication can be by audio alone or by audio with still images transmitted over telephone lines (AUDIOGRAPHICS). Video interaction can also be accomplished by FIBER OPTICS cable using DIGITAL COMMUNICATION technology operated by the established telephone systems. More often, the linkup connects at least some of the participants via SATELLITE. The transmission can be from one point to another point or from one point to multiple points. Usually the connection consists of two-way audio with one-way video. When the transmissions involve two-way audio and two-way video, most practitioners use the term "videoconferencing." (See also INTERNATIONAL TELECONFERENCING ASSOCIATION [ITCA]).

telecourse   an INSTRUCTIONAL TELEVISION [ITV]) course that is designed for secondary or college students and adults. The term is seldom used to describe supplementary/enrichment programs at the elementary or middle school level.

Telecourse lessons are sequential in nature and develop into a series by building upon one another in a linear fashion. Each lesson, however, is usually self-contained. The students combine television viewing with correspondence study, reading, self-instruction, and occasional on-campus sessions, along with communication with the teacher via telephone, computer, and fax machine. Courses are offered for credit and noncredit. The television portion of the course usually consists of 26 half-hour lessons. This normally meets the required number of student contact hours in a college or university if each half-hour contains the amount of information equivalent to that included in a 90-minute regular classroom lecture.

Telecourses are transmitted by cable, by PUBLIC TELEVISION (PTV) stations, on INSTRUCTIONAL TELEVISION FIXED SERVICE (ITFS) operations, by CLOSED CIRCUIT TELEVISION (CCTV) on a campus, and (increasingly) via SATELLITE in a DISTANCE EDUCATION situation. Some experiments have been conducted using VIDEOTEXT and TELETEXT in combination with the courses. (See also ADULT LEARNING SERVICES, AG*SAT, ANNENBURG/CPB PROJECTS, MIND EXTENSION UNIVERSITY [MU/E], and OPEN UNIVERSITY.)

Telecine   See FILM CHAIN.

Telematique   a coordinated program by the French government to bring cable, computers, VIDEOTEXT, and TELETEXT to every French citizen by 1995. In addition to planning for FIBER OPTICS cable and addressable converter capabilities, the government has funded the creation and operation of the ANTIOPE version of videotext. The French government is committed to the development of these and other technologies as a national policy under the Telematique program.

telenovelas   Latin American versions of SOAP OPERAS. Often called *novelas,* the daily serials differ from the U.S. soaps in that they have a definite ending. While they can consist of more than 100 episodes, they eventually come to a conclusion in months, not years. The 15-minute and half-hour shows vary in style and content but most are about love and romance.

telephone coincidental survey   a technique that determines the number of viewers of a television program. Many researchers believe it to be the most accurate of all methods, because it can estimate the extent of the audience at a precise, given moment. Rating company employees phone sample households and ask them if and what they are watching at that time. The method is called coincidental because the questions and the call coincide with the activities underway in the household. The research, by its very nature, covers only one-quarter of an hour of viewing. The method is relatively expensive and it is not used on a regular basis by audience research firms such as A. C. NIELSEN. (See also OVERNIGHT RATINGS.)

telepics   See MADE-FOR-TV MOVIES.

teleplex   a new entity by which the local PUBLIC TELEVISION (PTV) station becomes a nucleus for exploiting all information, video, and communication technology for public purposes. The term replaces the unwieldy "public telecommunication center complex" term, according to its originator, James A. Fellow, president of the CENTRAL EDUCATIONAL NETWORK (CEN). The teleplex (from the Greek *tele* meaning far-off or distant and the Latin *plex* meaning linking) ". . . consolidates and coordinates the organization and management of various capacities associated with electronic communications," according to Fellows. Under this concept, the PTV station expands its broadcast role, thus applying all modern communication technology to information/educational as well as to entertainment purposes.

teleports   engineering complexes that contain on-line technical capa-

bilities primarily related to SATELLITE communications. Teleports provide UPLINK and DOWNLINK services for audio, video, and data and voice transmission on KU-BAND and C-BAND SATELLITES. They normally have MICROWAVE RELAY and FIBER OPTIC capabilities for technical interconnections to and from producers, stations, SYNDICATORS, and SATELLITE NEWS GATHERING (SNG) operations and can provide encryption (SCRAMBLING) services. Some teleports also have production and POST-PRODUCTION facilities.

TelePrompters   electronic devices that assist the talent on a television set in delivering the words from a script. A small closed-circuit camera feeds a moving image of the written script to monitors below the camera lens. A double mirror system, which allows the reader to look directly at the lens and still see the script, permits the talent to maintain that all-important eye contact with the viewing audience. (See also CUE CARDS.)

telestrator   See STOP ACTION.

Teletel   a two-way French VIDEOTEXT system that has struggled to become a national service. It is sometimes referred to as "Minitel." The system received a great deal of attention for its development of an electronic phone book. An individual can query a computer data base for a telephone number, which appears on the television screen. Other uses involve "smart cards," which allow the user to insert a slim credit card into a home terminal to transfer money or pay bills. In addition, families can shop at home, make hotel reservations, and participate in COMPUTER-AIDED INSTRUCTION (CAI), using the home computer keyboard, television set and the connection to the main terminal. (See also ANTIOPE and TELEMATIQUE.)

teletext an information transmission system that delivers print and graphics via broadcasting to television sets. It was authorized by the FEDERAL COMMUNICATIONS COMMISSION (FCC) in 1983. The system does not send moving pictures, and it is primarily a one-way method of transmission, as opposed to VIDEOTEXT, which allows for two-way communication over telephone lines or cable. In the United Kingdom, both teletext and videotext are often called by the generic term "viewdata."

In a teletext system the information is sent out over the airwaves using a part of the television station's signal. The signal is received at home where the vertical BLANKING interval captures it. The information can then be retrieved by the viewer using a hand-held DECODER.

The information can be viewed separately from the regular television picture, or both the television programming and digital information can appear on the screen at the same time, a process used in CLOSED CAPTIONING. Teletext is a modest form of INTERACTIVE TELEVISION because the viewer can select the type of data that is desired. (See also ANTIOPE, CEEFAX, ELECTRA, NORTH AMERICAN BROADCAST TELE-

TEXT STANDARD [NABTS], ORACLE, TELIDON, VIDEOWAY, and WORLD STANDARD TELETEXT [WST].)

television according to the FEDERAL COMMUNICATIONS COMMISSION (FCC): "Television is the synchronous transmission of visual and aural signals. The picture phase is accomplished by sending a rapid succession of electronic impulses, which the receiver transforms into scenes and images." The accompanying audio is transmitted separately but simultaneously via FREQUENCY MODULATION (FM) broadcasting. The term is a combination of Greek and Latin roots. *"Tele"* is Greek for "far-off" and *"vision"* is derived from the Latin verb *videre* (to see).

Television Bureau of Advertising (TvB) a marketing service organization that conducts research and supplies industry information for use at local stations and at the networks to help sell COMMERCIAL TIME. The bureau promotes broadcast television as an advertising vehicle and seeks to increase television's share of the advertising dollar. (See also CABLETELEVISION ADVERTISING BUREAU [CAB].)

Television Code a now-defunct series of recommended standards for television advertising and programming developed by the NATIONAL ASSOCIATION OF BROADCASTERS (NAB) in 1952. The code was in effect for years and at the height of its influence, was subscribed to by some 60 percent of the commercial television stations in the United States. It forbade "profanity, obscenity, smut, and vulgarity" in programming, and sex crimes were not deemed suitable topics for broadcasting. In 1975 (under pressure from the FEDERAL COMMUNICATIONS COMMISSION (FCC) the code embraced the concept of a FAMILY VIEWING TIME. The most specific recommendations of the code were concerned with the subject matter and the time limits of COMMERCIALS. It also forbade liquor commercials and such techniques as SUBLIMINAL ADVERTISING.

The NAB stressed that the code was voluntary, but the Justice Department believed that stations were coerced into adhering to it and brought suit against it in 1979 on antitrust grounds. After a number of legal maneuvers, a federal district court approved a consent decree in 1982 in which the NAB agreed to drop the code.

Television Critics Association (TCA) a national organization of newspaper and magazine critics of television programming. It organizes an annual meeting in Hollywood each July where the networks, cable services, and SYNDICATORS present their fall programming.

Television Decoder Circuitry Act of 1990 an amendment to the COMMUNICATIONS ACT OF 1934 passed by Congress in 1990, which required that all new television sets sold or imported in the United States after July 1, 1993 with a picture screen 13 inches or greater be equipped with a built-in CLOSED-CAPTIONED DECODER. The FEDERAL COMMUNICATIONS COMMISSION (FCC) was assigned the responsibility of enforcing the law, which was designed to expand the accessibility of closed-captioned technology. (See also NATIONAL CAPTIONING INSTITUTE [NCI].)

television receive only (TVRO) antennas saucer-shaped ANTENNAS that allow home owners to receive television signals directly from a SATELLITE rather than from a cable system or from a terrestrial television broadcast station. The backyard dishes range in size from six feet to 15 feet in diameter and are sometimes known as "C-BAND direct antennas" because they are tuned to pick up signals from those satellites. They are used in what has become an early form of a DIRECT BROADCAST SATELLITE (DBS) system.

Telidon a TELETEXT and VIDEOTEXT system for digital character coding, which can be transmitted over broadcast, telephone, or cable systems. It was developed by the CANADIAN DEPARTMENT OF COMMUNICATIONS and operates from a sophisticated computer data base. The user can access a wide variety of information by a terminal and a computer keyboard.

Telstar satellite an experimental communication SATELLITE that relayed the first video image over the Atlantic in 1962. The picture was a live shot of an American flag on a pole outside of the transmitting station in Maine.

tent   card   a   point-of-purchase (POP) merchandising device that is a small advertising display for an item in a retail store or home video outlet. It is imprinted on two sides and folded so that it can stand alone and be read from either side.

tent pole a television programming term describing a show that sits in the middle of the PRIME TIME schedule. It supports the evening programming on a network, and the other shows are scheduled around it.

terrain shielding a broadcasting situation in which mountainous or other irregular terrain blocks or weakens the transmitted signals of a radio or television facility. Such topographical shielding sometimes prevents interference with the signals of other nearby broadcast operations.

test pattern an optical chart used to calibrate and align electronic equipment. It contains geometric patterns in circles and squares to allow technicians to align cameras and the output of the signals of other equipment for optimum resolution, focus, contrast, linearity, and framing in MONITORS and home receivers.

testimonial commercials specialized COMMERCIALS featuring individuals who endorse a product or service by indicating satisfaction with it. The testimonials can come from a number of ordinary people or from a celebrity.

three-D video an experimental technique that offers motion pictures for three-dimensional viewing on videocassette. A kit contains a stereo driver that is attached to the consumer's VCR and a set of electronic glasses that plug into the device. Looking at the screen with the glasses on during the playback creates a 3-D effect.

3/4-inch U (EIAJ) video recording format the first mass-produced and practical videocassette format and machine to be used in the United States. Introduced by Sony in 1971, this VIDEOTAPE FORMAT uses the trade name U-matic (shortened to just U), which has become synonymous with the machine. It was endorsed as the standard for 3/4-inch tape recording by the Electronics Industries Association of Japan (EIAJ). The machine has been extremely popular in the AUDIOVISUAL COMMUNICATIONS and CORPORATE TELEVISION markets.

THX   sound   a   new   consumer sound system designed to provide the best theater-quality sound for videocassette and videodisc viewing in the home. The system is designed to bring out the unrealized audio potential of theatrical film sound tracks and to handle soft passages, high trebles, and deep bass tones with superior clarity.

The system was developed by engineer Tomlinson Holman and filmmaker George Lucas in 1990, and the name refers to the first Lucas film *(THX1138)* or to "the Holman experiment," according to the inventors.

tiering (cable marketing) the group sale of various cable program services to consumers. The system bunches certain BASIC CABLE and PAY (PREMIUM) CABLE networks together, offering potential customers discounts if they subscribe to the various packages.

In the classic tiering method, customers are offered a basic service that consists of local television stations, some DISTANT SIGNALS, and one or more nonbroadcast services such as a community bulletin board. The subscriber can then order other channels at extra fees on an individual basis or as part of a tier (or block) of different channels. The actual contents of the various tiers varies with the MULTIPLE SYSTEM OPERATOR (MSO). Some group their offerings according to the audience, such as "family," "men," "women," and "kids" services. Others organize them by subjects: "shopping," "news/information," "music/entertainment," "arts/music," and "sports."

tiering (program) a method of buying a FIRST-RUN program. Stations negotiate deals in which they obtain the option of running the program in any program tier, such as EARLY FRINGE or LATE NIGHT. The station then pays an appropriate negotiated price based on where the show actually runs in the schedule.

tilt the rotation of a stationary camera on its vertical axis. The operator moves the camera up or down to follow the action, such as when people stand up or sit down. (See also FRAMING and PAN.)

time bank syndication a transaction whereby a national advertiser or its ADVERTISING AGENCY may provide free programs to a broadcast station or cable system in exchange for future COMMERCIAL TIME. The time is "banked" as SPOTS and is used when the agency or its client determines that it will best fit a particular advertising CAMPAIGN. This type of syndication is not as popular as the FULL BARTER, SPLIT-BARTER, or CASH-BARTER forms of syndication.

time base corrector (TBC) an electronic device used to correct synchronization (SYNC) or timing errors in videotape playback. The apparatus is used to correct the slight but common instability of a signal from different videotape machines. It removes JITTER, jumps, or rolls from the signal and corrects color, timing, and sync errors before passing the signal on to its next stage or destination.

time shift a buzz word used to promote the usefulness of home videocassette recorders (VCR). The term was coined by the chairman of the SONY CORPORATION. The machines were touted as being able to shift time by their ability to record a television program OFF THE AIR for playback at a later time.

time-lapse video recording a type of video recording that captures images over a very long period of time. Using a special videotape recorder (VTR) connected to a camera, the process can record anywhere from eight to 200 hours of action, depending on the VTR model. TImelapse recorders are used to capture scientific experiments and for surveillance and security purposes with HIDDEN CAMERAS in malls and banks.

Time Warner Inc. a megaglomerate created in 1989 by the acquisition of the many Warner Communications Inc. interests in television, cable, publishing, and music by Time Inc. The resulting firm became the world's largest media and entertainment company.

The corporation owns 25 magazines including *Time, Life, Fortune, Sports Illustrated, People,* and *Entertainment Weekly.* Time Warner Books is the second largest book publisher in the United States. The company also owns the imprints Little Brown and Time-Life Books and operates the Book-of-the-Month Club. Warner Bros. motion picture studio is a consistent producer of hit movies.

The corporation is involved in home video through Warner Home Video and in 1990 it became the largest worldwide video PROGRAM SUPPLIER with the purchase of the home video rights to MGM and UNITED ARTISTS (UA) films from MGM/PATHE. It is also a major producer and distributor of television programming, with WARNER BROS. TELEVISION PRODUCTION and Lorimar Television Production. WARNER BROS. DOMESTIC AND INTERNATIONAL TELEVISION DISTRIBUTION COMPANIES are among the largest SYNDICATORS of FIRST-RUN programming.

Time Warner is also involved in cable television, both as a programmer and an operator. It owns the nation's oldest PAY (PREMIUM) CABLE network, HOME BOX OFFICE (HBO), its sister Cinemax, and the newer COMEDY CENTRAL. The Time Warner Cable Group manages two large MULTIPLE SYSTEM OPERATORS (MSO), the AMERICAN TELEVISION AND COMMUNICATIONS CORPORATION (ATC) and WARNER CABLE COMMUNICATIONS.

total audience rating (TAR) an audience measurement rating that is an estimate of the percentage of the population in a market that has been tuned into a program for a minimum of five successive minutes. It also reflects the audience of a television station or a cable operation over a period of time. (See also AVERAGE AUDIENCE.)

tracking an electronic process causing the heads in a videotape playback device to exactly follow the tracks laid down in an original recording. Most often used in playing back a tape that has been rerecorded on a different (but same VIDEOTAPE FORMAT) machine, a tracking control device electronically compensates for the minor differences in the alignment of the recording heads between different videotape machines and corrects JITTER and SKEWING.

traction a programming term describing the condition of a show that has established itself with a small but regular audience. It attracts the same DEMOGRAPHICS each week and has the possibility of growth, having established some traction.

trade advertising   print advertising within an industry (for CONSUMER ELECTRONIC products or for television programs) that is designed to persuade retail stores to purchase the machines for resale to customers or to persuade stations to buy the programs to broadcast them to their audiences. SYNDICATORS do a considerable amount of this type of advertising, as do home video PROGRAM SUPPLIERS. The purpose is to promote and increase the outlets that will handle a product or program. The opposite of consumer advertising, trade advertising is usually confined to TRADE MAGAZINES.

trade association   a nonprofit organization made up of members from a particular industry. Membership can be on an individual basis or by company, organization, or institution. The memberships of the principal associations in the communications world are largely companies or institutions, because of the relatively high annual membership dues. Trade associations represent their members' interests before Congress, hold seminars, publish newsletters and magazines, and conduct regional and national membership meetings and trade shows.

trade magazine   a type of magazine covering a particular industry, with news and features about that field. Often called "business publications," their content is usually of interest only to people within the field.

Such publications ("books," in magazine parlance, and "the trades" or "trade press" within an industry) are often controlled-circulation periodicals that are sent free of charge to people within a particular segment of an industry. The magazine receives its revenue from TRADE ADVERTISING.

trade show   a periodic exhibition of products or services, usually of interest only to the professionals in a particular industry. Companies rent booth space in an exhibit hall to showcase their goods or services to others in that field. Most trade shows are annual affairs, sponsored by various nonprofit TRADE ASSOCIATIONS and are adjuncts to the national membership meetings of the groups. Many associations are primarily supported by the income received from the exhibitors at their trade shows.

trafficking rules   FEDERAL COMMUNICATIONS COMMISSION (FCC) rules that curbed some tendencies to unduly profit in the buying and selling of radio and TV stations. To forestall the practice of simply acquiring a CONSTRUCTION PERMIT (CP) or license with the sole interest of selling it for a quick profit, the Commission established antitrafficking rules. The rules stated that a CP could not be sold at any time for a profit. More important, they required an owner to operate a station for at least three years before the station and the license could be sold. The FCC repealed the three-year license rules in 1982 while retaining the CP rules.

trailers short videos used to promote a film or program on videocassette. Similar (and often identical) to the brief promotional films shown in motion picture theaters, they feature highlights of the title, including snappy dialogue, exciting sequences, and intriguing shots. They, in fact, take their name from such motion picture promos.

transistor a tiny device that provides the same oscillation, switching, and amplification functions of the vacuum tube, but is much smaller, more reliable, and requires less power. Consisting of semiconductors, it made possible the development of compact consumer, PROSUMER, and professional electronic equipment. It is capable of $1000 \times$ amplification. The term was devised by combining the two words "transfer" and "resistor."

translator station a low-power broadcasting operation. It receives a signal OFF THE AIR from a primary full-power television station, converts (or translates) the incoming signal to another frequency, amplifies it, and retransmits it to the public. Translators using UHF or VHF frequencies expand the coverage of conventional broadcast stations. Most translators are located in rural areas. They are particularly useful in the western plain and mountain areas of the United States where their low cost and 15-to-20-mile radius can extend the coverage of major stations, bringing television to remote and rural areas. Translators can serve as LOW POWER TELEVISION (LPTV) stations and many have been converted to and authorized as such by the FEDERAL COMMUNICATIONS COMMISSION (FCC). (See also BOOSTER and SATELLITE STATIONS.)

transmitter a device that receives incoming audio and video signals from a source (videotape recording, live, etc.) and feeds a composite signal to a diplexer and then to an ANTENNA for redistribution through the airwaves. The transmitter is usually located in a building adjacent to the tower and antenna, and the source of the incoming signal is often a STUDIO-TRANSMITTER LINK (STL). The transmitter uses the audio/video signal to modulate a carrier wave at the station's assigned frequency. The transmitter is authorized by the FEDERAL COMMUNICATIONS COMMISSION (FCC) to operate at a certain level of radiated power expressed in watts. (See also EFFECTIVE RADIATED POWER and MODULATION.)

transponder an electronic component of a communication SATELLITE that receives and translates the audio and video signal from an UPLINK to another FREQUENCY and retransmits (DOWNLINKS) the signal back to Earth. Satellites contain up to 24 transponders, each capable of receiving and transmitting one channel.

traps electronic devices used by cable systems to prevent unauthorized (and unpaid) reception of cable channels by subscribers. They are installed between the MULTITAP on the FEEDER LINE and the CABLE DROP line going to the subscriber's home. Because they are usually attached to the feeder line above the ground near a telephone pole, they are difficult to tamper with and are a relatively effective antidote to PIRACY. (See also SCRAMBLING.)

tree network a cable distribution system in which one origination point transmits a signal that, through various branches, finally reaches subscribers in their homes. The design resembles a tree inasmuch as the signals from the HEADEND (the root) are carried to TRUNK LINES (major limbs) and then through FEEDER CABLES (large branches) to CABLE DROPS (small branches), and finally to the individual converter (leaf) in the home. (See also HUB SYSTEM and STAR SYSTEM.)

tripod a three legged mounting device that supports a head on which a camera is mounted. Most tripods have an adjustable middle vertical pedestal, which can be cranked up or down to raise or lower the camera mounting head and thus the camera. The tripod legs also often telescope so that the camera can be raised or lowered by as much as three feet.

tristandard videocassette recorder a videocassette machine that can record and play back in all three of the major television standards (SECAM, PAL, and NTSC). A tristandard MONITOR must be used to display the images.

truck a camera movement in which the camera mount is moved left or right at the command of the director to "truck right (or left)." The effect is a sideways parallel motion, past the subject. It is sometimes called a "crab." The term is also used in television to refer to a remote truck. (See also ARC and FRAMING.)

trunk lines the main highways of a cable operation in a TREE NETWORK. The trunk lines begin the distribution of the electronic signals from the HEADEND to the subscriber's TV set. Consisting of large COAXIAL CABLES, three-quarters to one inch in diameter, they carry the signal to smaller FEEDER CABLES, which continue the distribution to even smaller CABLE DROP lines, which connect to the subscriber's home. BRIDGING AMPLIFIERS are placed at periodic intervals (every one-third to one-half mile on trunk lines) to boost the original signal and correct the problem of ATTENUATION.

tune-in advertising a form of print promotion by a television or cable operation that encourages the viewer to watch a particular program. Typically placed in newspapers on the same day that the program airs, such advertising alerts the reader to the upcoming program. The corollaries of tune-in advertising in the electronic media are on-air PROMOS.

Turner Broadcasting System Inc. (TBS) a corporation with a major leadership role in communications in the United States. Starting with the independent UHF television station WJRJ in Atlanta (now WTBS

SUPERSTATION), Ted Turner built a corporation that has reshaped the television industry. Turner purchased the station in 1970 and by 1976, he had originated the superstation concept, beaming the station's signal via satellite-to-cable systems throughout the country. In 1980 Turner inaugurated the CABLE NEWS NETWORK (CNN), the first round-the-clock all-news network. A second live all-news service, Headline News, began operation in 1982. In 1986 TBS acquired the MGM Entertainment Company, which had assets that included a library of feature films and the MGM theatrical motion picture and television production business. The library was used to form the cornerstone of Turner Network Television (TNT), another BASIC CABLE network, which was launched in 1988. Major subsidiaries of Turner Broadcasting include Turner Program Services, formed in 1981 as the company's SYNDICATION arm, and Turner Home Entertainment, inaugurated in 1987 to serve the home video market.

turnover   a term with applications in both television and home video. In television advertising and programming, it refers to the audience tuning in and out of a given program. In the home video industry, each rental transaction is referred to as a "turn." A high turnover rate on a popular title means that the retailer will receive a fast rate of return on his investment in that title. The word "turnover" is also used as a description of the rate at which INVENTORY is sold or turned over.

In Europe, turnover often refers to the cash flow of a business.

TV Marti   a TV station that transmits American-produced programming to Cuba. It is a project of the UNITED STATES INFORMATION AGENCY (USIA). The service was modeled on Radio Marti, an AM station that began broadcasting programming to Cuba in 1985. TV Marti was inaugurated in 1990 after considerable debate in Congress over its feasibility, cost, and probable effectiveness.

TV Ontario   a nonprofit Canadian entity that is the largest producer and distributor of INSTRUCTIONAL TELEVISION (ITV) programs in the world.

The organization is engaged in broadcasting, cable, and video programming in both English and French. It acquires K–12 and college TELECOURSES and adult programs, but the vast majority of its programs are produced in-house. The group distributes its programs to schools, colleges, and PUBLIC TELEVISION (PTV) stations throughout the United States and in various other countries. Many of the programs are used for DISTANCE EDUCATION purposes.

TVQ rating   a qualitative audience research tool that measures the "likability" and familiarity of performers, sports figures, and animated or nonhuman properties in the entertainment world. The ratings are used by producers and ADVERTISING AGENCIES in developing and casting programs. The personalities, shows, or products are rated according to their appeal as well as the extent to which they are well known or easily recognized.

TVRO See TELEVISION RECEIVE ONLY (TVRO) ANTENNA.

12-and-25-percent rule a FEDERAL COMMUNICATIONS COMMISSION (FCC) rule regulating the extent of ownership of broadcast stations. It became effective in 1985, replacing the 777 RULE that for more than 30 years had limited ownership of broadcast stations to seven of each type. The new rule increased the number of AM and FM radio stations and television stations that a single entity could own from seven to 12 of each type. In addition the FCC limited a single company's total audience potential to 25 percent of the television households. In recognition of the technical disparity between VHF and UHF, the FCC rule allows a 50 percent "discount" of the percentage of audience served by UHF stations toward the 25 percent television household ceiling.

In 1992 the FCC moved to further relax ownership rules by permitting a single company to own as many as 30 AM and 30 FM radio stations. The 12-station limit remains for television stations. (See also DUOPOLY RULES.)

two-step distribution a process of moving goods from the manufacturer to the consumer, traditional in the prerecorded video industry. The videocassette or videodisc passes from the PROGRAM SUPPLIER through the two steps of the WHOLESALER and retailer to the consumer. In contrast, most of the record industry follows a one-step pattern whereby the retailer buys directly from the LABEL/manufacturer.

type B videotape format the most widely used VIDEOTAPE FORMAT in Europe. It is also popular among professionals in other parts of the world. The machine uses 1-inch reel-to-reel videotape and produces high quality recordings. It, along with the TYPE C VIDEOTAPE FORMAT that is popular in the United States, has largely replaced the older 2-inch QUADRUPLEX (QUAD) VIDEOTAPE RECORDING machines.

type C videotape format a VIDEOTAPE FORMAT that is widely used in network and broadcast operations in the United States. The 1-inch reel-to-reel machine creates superior recordings and (along with the TYPE B machine) has replaced the older 2-inch QUADRUPLEX (QUAD) VIDEOTAPE RECORDING machines in professional circles.

U

UHF ultra high frequency. The initials designate the electromagnetic BAND from 470 to 806 MEGAHERTZ (MHZ) that contains television channels 14 to 69. These channels were added to the television spectrum by the FEDERAL COMMUNICATIONS COMMISSION (FCC) in 1952 in its historic SIXTH REPORT AND ORDER. The new UHF channels were added to the small number of existing channels (12) in the VHF band. The eventual result was to make possible many more television stations in the United States. (See also ALLOCATION, THE FREEZE, FREQUENCY, and UHF TELEVISION STATIONS.)

UHF television stations television stations operating on channels between 14 and 69 in the United States, Canada, and some countries in western Europe. They are known in the trade as Us. They were made possible by the SIXTH REPORT AND ORDER of the FEDERAL COMMUNICATIONS COMMISSION (FCC) in 1952, which allocated 2,053 channels in the broadcast spectrum to geographically separated markets throughout the United States. Some 66 percent (or nearly 1,400 channels) were in the UHF band. The reasons for adding the UHF channels to the existing 12 channels in the VHF (very high frequency) band was to increase the total number of television stations throughout the country. (See also ALL-CHANNEL LAW.)

UNDA–USA an organization that is the American branch of the International Catholic Association for Radio and Television. The name of the group was taken from the Latin word "unda" meaning "wave" (in this case, airwave). UNDA–USA members include Catholic broadcasters, dioceses, syndicated programmers, radio and television stations, and associated agencies. (See also GABRIEL awards.)

underground television See ALTERNATIVE TELEVISION.

underwriting the private financial support of programming on noncommercial television. The PUBLIC TELEVISION (PTV) stations are forbidden by the FEDERAL COMMUNICATIONS COMMISSION (FCC) to broadcast COMMERCIALS or accept SPONSORS for particular programs. The financially strapped industry, however, encourages the donation of funds to support (underwrite) the production of programs or the operation of the stations. The producer or distributor often solicits financial support from companies, foundations, or organizations to produce a specific series or program. In return for such support the organization receives a brief credit at the beginning and end of the program when it is aired. Because all PTV stations are nonprofit entities such gifts also provide tax advantages for the donating companies. The FCC has established underwrit-

ing rules. (*See also* PUBLIC BROAD-
CASTING SERVICE [PBS]).

UNESCO   the United Nations Educational, Scientific, and Cultural
Organization—a branch of the
United Nations. UNESCO utilizes
AUDIOVISUAL COMMUNICATIONS and
media in its primary mission of
promoting freedom and human
rights through cooperative international efforts. With 160 member
nations the agency provides information and books and assists countries in equipping themselves with
media including radio, EDUCATIONAL
TELEVISION (ETV), film, and video.
UNESCO has become increasingly
politicized, and the United States,
which was the largest contributor of
funds to the agency, withdrew from
the organization in 1984. (*See also*
INTERNATIONAL COUNCIL FOR EDUCA-
TIONAL MEDIA and NEW WORLD
INFORMATION AND COMMUNICATION
ORDER.)

unidirectional microphone   an extremely sensitive microphone with a
pickup pattern in the one (uni) direction in which it is pointed. It is relatively insensitive to sounds coming
from the sides or rear. (*See also*
CARDIOID MICROPHONES and SHOT-
GUN MICROPHONES.)

United Artists Entertainment (UA)
the nation's largest motion picture
theater chain, which is also one of
the largest cable MULTIPLE SYSTEM
OPERATORS (MSO). The company was
initially formed as a merger of the

cable operator United Cable and
United Artists Entertainment and
grew with the acquisition of cable
systems. In 1991 UA merged with
and became a subsidiary of Telecommunications Inc. (TCI).

**United Church of Christ Office of
Communication**   a media action
organization whose purpose is to
help citizen groups gain access to
television and to improve the media.
The office is dedicated to promoting
citizens' rights and it has been particularly effective in advocating minority concerns in the media.

**United Nations Correspondents
Association (UNCA)**   a nonprofit
association of professionals who are
accredited U.N. press, radio, and
television news reporters. The organization works to maintain the freedom of the press in relationship to
the United Nations.

**United States Information Agency
(USIA)**   an independent foreign affairs agency within the executive
branch of the federal government. It
explains and supports U.S. foreign
policy and national security interests
abroad through a number of information programs. The agency maintains 205 posts in 128 countries,
where it is known as USIS, the U.S.
Information Service. The agency's
purpose is to increase mutual understanding between the people of the
United States and the people of
other countries. It is prohibited from
dissemination within the United
States of any materials produced by

the agency for distribution overseas. (See also TV MARTI and WORLDNET.)

United States Satellite Broadcasting (USSB) a DIRECT BROADCAST SATELLITE (DBS) company. In 1994 customers will be able to see its programming with their 15-inch TELEVISION RECEIVE ONLY (TVRO) dishes.

universe the total population or audience that may have some interest in a program, product, or service. Using DEMOGRAPHIC and PSYCHOGRAPHIC data, researchers break down such a universe into a TARGET AUDIENCE. The term is also used in home video direct mail campaigns in referring to the mailing list from which sampling is done.

University Consortium for Instructional Development and Technology (UCIDT) a nonprofit organization composed of colleges and universities with research and educational programs in AUDIOVISUAL COMMUNICATIONS and technology. Its purpose is to improve graduate studies in media and to foster professional standards.

University Film and Video Association (UFVA) a nonprofit organization consisting of student video and filmmakers at colleges and universities as well as teachers who are involved in the production and study of film and video. The organization sponsors research programs, maintains a placement service, and awards scholarships and grants.

University Film and Video Foundation a private foundation that supports college and university film and video productions. The foundation helps distribute independent film and video productions and awards small scholarships and fellowships.

university stations types of PUBLIC TELEVISION (PTV) stations licensed by the FEDERAL COMMUNICATIONS COMMISSION (FCC) to colleges or universities throughout the United States. They are the second-most-numerous type of PTV stations.

Univision a company that provides a full range of Spanish-language programming produced both in the United States and by communication companies throughout the Spanish-speaking world. The firm transmits movies, TELENOVELAS, sporting events, variety shows and newscasts entirely in Spanish to cable systems and stations. The New York-based GROUP BROADCASTER is owned by Hallmark Cards Inc. and First Chicago Venture Capital. It is considered the dominant Spanish-language network in the United States.

unwired network a temporary assemblage of television stations airing the same commercials. The practice is a hybrid of a MARKET-BY-MARKET BUY and a NETWORK BUY. Companies purchase AVAILABILITIES at a discount from a number of stations and then sell them as a group to national advertisers. Most unwireds sell most of the time in UPFRONT BUYS and the rest in the SCATTER MARKET. They offer MAKEGOODS if the RATINGS (as measured by PEOPLE METERS) do not reach specified goals. Unwired rates are lower than those for regular spots.

upfront buy sales of COMMERCIAL TIME to advertisers during the summer months for the fourth quarter and the first three quarters of the following year. National advertisers make these upfront buys of commercial time at lower prices than they might get during the later SCATTER MARKET. The purchases are based on a guarantee that the commercial will reach a specific number of viewers when it is shown later in the program year. If it does not, the network or SYNDICATOR must usually provide MAKEGOODS whereby the commercial is repeated.

upgrade the addition of new or BASIC CABLE or PAY (PREMIUM) SERVICES to the subscriber's home. It is the opposite of DOWNGRADE. In cable engineering, the expression refers to a major physical improvement in the system. New electronic and COAXIAL CABLE components are added to the operation in order to increase signal capacity. An upgrade, however, is not as extensive as a REBUILD.

uplink the entire ground-to-sky SATELLITE system. It includes the terrestrial TRANSMITTER and ANTENNA and associated electronic equipment of the EARTH STATION as well as the receiving TRANSPONDERS on the satellites. The term is often specifically and erroneously applied only to the large concave dish used to transmit the signal. The term is also used (correctly) to describe the entire process in which a signal is "uplinked" to a satellite. (See also DOWNLINK and TELEPORTS.)

upscale people at the top end of a DEMOGRAPHIC scale or products of class and distinction. People with high levels of education and income or members of the professions are considered to be upscale and become a TARGET AUDIENCE for expensive products or services. Programs, products, and COMMERCIALS are then developed to appeal to their tastes. The converse term, "downscale," is seldom used in advertising or programming circles because of its negative connotation.

V

value-added strategy a technique used in selling video hardware and software in which the retailer offers an additional item or items free of charge as an incentive to the customer. For example, two movie tickets are included in the sale price of a prerecorded videocassette. The goal of the value-added technique is to convince the hesitant customer to make a purchase by using what is considered to be the ultimate argument.

Vanderbilt Television News Archive an archive housing videotapes of the nightly news broadcasts of NBC, CBS, ABC, and one hour of the broadcast day of the CABLE NEWS NETWORK (CNN). It contains thousands of hours of news programs with additional hours added each week. The programs are taped off the air from local stations and cable systems in Nashville. Tapes in the archive are available for study on the premises or through rental.

Varietese a unique show-business language, originated by the TRADE MAGAZINE *Variety.* Its reporters have coined many colorful and descriptive words and phrases that have become standard industry jargon. Among them are SITCOM, anchor, EMCEE, tel-epic (for a MADE-FOR-TV MOVIE) and (in the 1930s) SOAP OPERA. The phrases are sometimes combined in intriguing headlines such as the famous "Stix Nix Hix Pix," which appeared over a story about the reluctance of rural audiences to attend movies depicting country life.

vast wasteland a description of commercial television in 1961 by the then newly appointed chairman of the FEDERAL COMMUNICATIONS COMMISSION (FCC), Newton Minow. It shocked and alarmed his audience

of broadcasters at the annual convention of the NATIONAL ASSOCIATION OF BROADCASTERS (NAB). He invited members of the audience to watch their own stations for one full day. "I can assure you," he said, "that you will observe a vast wasteland." The phrase was inspired by a poem by T. S. Eliot.

VCR Plus a small, remote-controlled battery-operated device that triggers the recording of a program OFF THE AIR or from a cable system on a videocassette recorder (VCR) when the viewer punches in a specific code on its key pad. The code corresponds to a given program and is published in the local newspaper or *TV Guide,* next to the program listing. The hand-held 4-by-6-by 1-inch gadget is designed to simplify the off-air recording of programs.

VCR-2 a dual deck videocassette machine consisting of two videocassette recorders (VCR). It allows convenient DUBBING and VIDEOTAPE EDITING for the home video buff. The patented device has two loading slots, side by side on the front panel. A blank tape is inserted in one slot and the tape to be copied in the other. Punching a button marked "Copy Tape" starts both cassettes rolling.

vectorscope a specialized, stand alone, electronic testing device that measures the purity of a color signal's hue and chrominance. Like an OSCILLOSCOPE or WAVEFORM MONITOR, the apparatus takes a small portion of the color signal generated by a camera or test generator and graphically displays the COLOR BARS continuously on a small round screen. The bars are represented, however, by six small boxes on the screen rather than by vertical bars. The vectorscope is more accurate than the human eye in measuring the correct balance in a color picture.

VH-1 a BASIC CABLE service that programs music videos that appeal to the 25-to-35-year-old age group. Programming to an older audience than the adolescents who view MTV, the channel features comedy routines, artists' SPECIALS, and some original musical programs, in addition to middle-of-the-road soft rock/pop music videos.

VHF very high frequency. The term designates the BANDWIDTH in the electromagnetic spectrum from 54 to 72 MHZ (channels 2 to 4), 76 to 88 MHz (channels 5 and 6), and 174 to 216 MHz (channels 7 to 13). These were the original channels allocated by the FEDERAL COMMUNICATIONS COMMISSION (FCC) to accommodate television broadcasting in the United States. The 12 channels (2 to 13) were the only ones available for use by broadcasters for television until the FCC issued its SIXTH REPORT AND ORDER in 1952, which added UHF channels to the system. (See also ALLOCATION, THE FREEZE, and FREQUENCY.)

VHF television stations television stations operating on channels between 2 and 13 in the United States, Canada, and some countries in western Europe. They are often called Vs in the trade. The stations were made possible when the FEDERAL COMMUNICATIONS COMMISSION (FCC) authorized commercial television broadcasting in 1941, when it adopted the NTSC black-and-white television standards. The activation of stations, however, was curtailed by both economics and World War II. The invention of the IMAGE ORTHICON TUBE (IO) and COAXIAL CABLE interconnection between some cities eventually spurred construction. By the 1960s, however, more UHF stations were in existence, many operating as INDEPENDENT STATIONS or EDUCATIONAL TELEVISION (ETV) stations. With the passage of the ALL-CHANNEL LAW in 1962 and the growth of cable television in the 1970s, the dominance of VHF stations came to an end.

VHS format a half-inch VIDEOTAPE FORMAT that has become the most popular in the United States. It dominates the home market and is also used in AUDIOVISUAL COMMUNICATIONS and CORPORATE TELEVISION. The initials stand for video home system. Developed by a subsidiary of MATSUSHITA and introduced in 1976, the videocassette machine won the format war for the loyalty of the American consumer 10 years later, even though most technical observers believed that it was inferior to the BETA FORMAT. The recording technology is also used in professional and consumer CAMCORDERS. (See also COMPONENT VIDEO SYSTEM/RECORDING.)

Viacom Enterprises one of the largest multimedia companies in the world. Beginning as a SYNDICATION firm, it distributed series, MINISERIES, feature films, and MADE-FOR-TV MOVIES from a vast program library. In 1994, the company won a five-month fight with QVC NETWORK INC, to acquire PARAMOUNT COMMUNICATIONS INC, and its many subsidiary firms. A planned merger with BLOCKBUSTER VIDEO will make the company the second-largest media conglomerate in the United States

video (definition) any visual image that can normally be seen by the human eye. The term is derived from the Latin "to see." In television engineering video is that portion of the signal that appears on the screen from the camera pick-up tube and sends it to the cathode ray tube (CRT), which displays it.

video compression See DIGITAL VIDEO COMPRESSION.

video high density (VHD) videodisc a videodisc that was heavily promoted in the late 1970s and early 1980s. It competed for press attention with the CAPACITANCE ELECTRONIC DISC (CED) and the LASER VIDEODISC (LV). The VHD technology was developed by companies in Japan and successfully introduced there. The VHD combined the best features of the LV and CED systems with a one-hour playing time on each side. Similar to the CED machine with a capacitance encoding system, the machine had a stylus that rode on the surface of a grooveless disc.

The disc itself was 10 inches in diameter and enclosed in a caddy.

Faced with the increasing popularity of the videocassette machine (which could record as well as play back) in the early 1980s, along with the counter claims of the two rival videodisc machines (which further confused the consumer), the Japanese companies decided against its introduction in the United States.

video magazines magazine-type information on a videocassette, periodically released to customers on a subscription basis. The idea has yet to gain wide acceptance from either consumers or advertisers. Subscribers are charged an annual fee (usually much more than a subscription to a printed magazine) and receive the videocassette through the mail.

video retail chain a number of home video stores in different geographic locations that are owned by one corporation. Strictly speaking, only two stores are needed to constitute a chain, but the term is usually applied to a group of at least 10 retail outlets. The chains offer a considerable challenge to the independently and singly operated MOM-AND-POP VIDEO STORES. Video retail chains vary in size from the mammoth national Blockbuster operation to smaller regional operations. Many of the chains operate VIDEO SUPERSTORES.

Video Retailers Association (VRA) a trade association formed in 1981 that consisted of video store owners from the New York area and later expanded to include a few members from Southern California and Chicago. The VRA failed in its attempt to affiliate with the National Association of Retail Dealers of America (NARDA), and the VRA and the VIDEO SOFTWARE DEALERS ASSOCIATION (VSDA) united under the VSDA banner in the summer of 1982 at the Consumer Electronics Show in Chicago.

Video Software Dealers Association (VSDA) the nonprofit TRADE ASSOCIATION of home video retail stores in the United States. The membership consists largely of the small and midsize independent MOM-AND-POP VIDEO STORES (rather than stores belonging to VIDEO RETAIL CHAINS) and the association reflects the members' interests and concerns. Associate members include PROGRAM SUPPLIERS and WHOLESALERS of prerecorded videocassette and videodisc programs and films, as well as companies that sell other products related to the industry.

video superstore large home video stores of at least 6,000 square feet of retail space, having an INVENTORY of 10,000 or more prerecorded titles for sale or rent. The outlets are usually owned by or operated under a FRANCHISE from a national or regional VIDEO RETAIL CHAIN. They attract customers with their supermarket size, clean, well-lighted atmosphere and the BREADTH AND DEPTH of their stock. They offer stiff competition to the smaller MOM-AND-POP VIDEO STORES.

videocassette See AUTOMATED VIDEOCASSETTE SYSTEMS, DESKTOP VIDEO, 8MM VIDEO FORMAT, HELICAL-SCAN VIDEOTAPE RECORDING, 3/4-INCH U (EIAJ) VIDEO RECORDING FORMAT, TRI-STANDARD VIDEOCASSETTE RECORDER, and VIDEOTAPE FORMATS.

videoconferencing See TELECONERENCING.

videodisc See CAPACITANCE ELECTRONIC DISC (CED), CD-I, CD-ROM, COMPACT DISC (CD), DVI, INTERACTIVE MULTIMEDIA, INTERACTIVE VIDEO, LASER VIDEODISC (LV), LASERDISC ASSOCIATION, VIDEO HIGH DENSITY (VHD) VIDEO DISC, and WORM.

videophiles a subsection of the people who are TECHTHUSIASTS. Videophiles are devoted to video cameras and production gear, and they purchase the latest equipment for use in amateur productions. They are sometimes called PROSUMERS.

videophonics the technique of sending and receiving pictures by telephone wire. Although TELECONERENCING allows individuals at one site to see and talk with people at another location, the technique is too expensive for day-to-day personal use. In 1988 the "picturephone" was introduced into the United States by Japanese firms. The videophones provide small, monochrome, still images of the users over regular telephone lines, using DIGITAL COMMUNICATIONS technology. Each image takes about five seconds to transmit and no one can talk during that time. Videophonics has failed to entice American consumers, however, and although the units are still sold in Japan they are no longer marketed in the United States.

videotape editing the assembling of television shots into sequences, sequences into segments, and segments into a finished program. In the process, good portions replace poor and are moved around into a final structured organization. The shots, segments, and sequences are joined electronically.

There are two basic strategies: assemble editing and insert editing. Assemble edits are accomplished in chronological order from beginning to end by assembling the title, then the first scene, second scene, third scene, etc., on a blank tape. Insert editing involves the replacement of an already existing shot, segment, or sequence with another. Insert edits are made to correct mistakes or to improve the program by inserting new and better material over old. Many editing projects begin with assemble editing in an OFFLINE EDITING mode and progress to insert editing using ONLINE EDITING techniques.

Videotape Facilities Association
See INTERNATIONAL TELEPRODUCTION SOCIETY (ITS).

videotape formats three major techniques developed for recording sound and pictures on magnetic tape: LONGITUDINAL (LVR), QUADRUPLEX (QUAD), and HELICAL-SCAN VIDEOTAPE RECORDING. LVR principles

are seldom used in video recording today, and the quad method is obsolete. The helical-scan technology has become the predominant method. The various formats are largely defined by the width of the magnetic tape used in recording. Generally speaking, a wider tape will produce higher quality images, because more electronic impulses can be embedded on the tape. Tape speed and other technical differences, however, must also be considered in defining formats.

The basic principle is that any tape recorded on one format should play back on any machine using the same format, regardless of the manufacturer of the machines, provided that the playback machine can operate at the same speed at which the tape was recorded. Conversely, a tape recorded in one format will not play back adequately on a machine using another format, even if they are both made by the same manufacturer and are operating at the same speed. In addition a tape recorded on a videocassette will not play back on a reel-to-reel machine.

Videotape Production Association
See INTERNATIONAL TELEPRODUCTION SOCIETY (ITS).

videotext an information system offering two-way electronic communication between a data base source and a home or business. It is a form of INTERACTIVE TELEVISION. Videotext differs from its one-way sister, TELETEXT, in that videotext information is transmitted via telephone line or cable rather than by broadcasting, and the subscriber can query the data base. The transmitting source is usually a computer containing a larger data base, which can be accessed by the subscriber via a special DESCRAMBLER/DECODER system and keyboard. The digital and graphic information is displayed on a regular television screen.

Videotext systems differ from database telephone line operations as they usually use a cable company's COAXIAL CABLE, and the home viewer selects the information desired using a keypad rather than a personal computer. The information is stored in a computer at the cable system HEADEND and/or the cable company subscribes to national data banks and simply acts as a relay to the cable subscriber. In the United Kingdom, the term is often spelled "videotex" but there, both teletext and videotex are often called by the generic term "viewdata." (See also HOME SHOPPING NETWORKS I AND II, HYPERMEDIA, PRESTEL, QUBE, QVC NETWORK INC., and VIDEOWAY.)

Videotext Industry Association (VIA) a nonprofit organization whose membership consists of firms engaged in the development, manufacture, or sale of VIDEOTEXT or TELETEXT equipment. It seeks to encourage the growth of videotext and teletext and to inform people about the technology.

videowalls a number of large television MONITORS arranged in vertical and horizontal rows, forming a rect-

angle or a square. The monitors are separated by thin borders and display the same or different crisp, bright high-quality images. The total size and shape of the multiscreen unit is dependent on the number of picture modules used and their assembled configuration.

Videoway a Canadian interactive cable television system. The operation is more of an INTERACTIVE TELEVISION system than the traditional VIDEOTEXT operation inasmuch as it offers an opportunity for viewers to select optional visual images rather than simply alternative printed texts. In the Videoway operation, a television station in Montreal broadcasts a regular program of news or sports. A sister cable system carries that broadcast on one channel but also offers an alternative on three adjacent cable channels. Using an interactive device located on the top of the set, a viewer at home can elect to receive a longer version of particular news stories, choose among various camera angles covering a baseball game, or request an instant replay of the action.

vidicon tube the most common pickup tube used in cameras for CORPORATE TELEVISION and AUDIOVISUAL COMMUNICATIONS. Consisting of a cylinder six inches in length and one-half inch in diameter, the tube is inexpensive but produces a slightly grainy picture. It is not as sensitive as the SATICON TUBE, but vidicon cameras remain the workhorses of the nonbroadcast television industry.

viewdata See TELETEXT and VID-EOTEXT.

virtual vision a production device that allows the cameraperson to see behind, above, or to the left or right without turning the head. A pair of goggles attached to a camera includes a tiny color TV with a small mirror in front of it. The picture is projected from the mirror onto a vi-sor. The user can view the picture, or if the visor is clear, can see the live action on the set.

voice-over an announcer reading audio copy or narration off-camera. It is usually recorded and then played back over the visual portion of the production. (See OFF MIKE/OFF CAMERA.)

W

warm-up a brief bit of business with a studio audience that combines a casual chat with jokes and questions and answers. It is commonly used in the preproduction of GAME SHOWS and SITCOMS and can last anywhere from 10 minutes to a half hour. The star, one of the leads, or the producer "warms up" the audience by appearing on stage and chatting informally with them. This period prior to the show is designed to get the audience in a mood that is receptive to watching or participating in the program.

Warner Bros. Domestic and International Television Distribution Companies two divisions of TIME WARNER INC., that handle the domestic and international SYNDICATION of WARNER BROS. TELEVISION PRODUCTION and Lorimar Television Production programs. The firms distribute feature films and OFF-NETWORK and FIRST-RUN SYNDICATED PROGRAMS from the vast Warner Bros. and Lorimar libraries.

Warner Bros. Television Production a branch of the giant TIME WARNER INC., which is consistently among the top producers of network television programs and FIRST-RUN syndication. Warner Bros. was the first major film studio to produce for television. The company's programs are syndicated by the WARNER BROS. DOMESTIC AND INTERNATIONAL TELEVISION PRODUCTION companies.

Warner Cable Communications a MULTIPLE SYSTEM OPERATOR (MSO) with cable systems throughout the United States. In 1989 it became a part of the TIME WARNER Cable Group, which was formed to manage the new megaglomerate cable operations after the acquisition of Warner Communications Inc. by Time Inc.

The group includes the AMERICAN TELEVISION AND COMMUNICATIONS CORPORATION (ATC).

waveform monitor   an electronic testing device that measures the black-and-white information in a color picture and the synchronization (SYNC) of a television signal. A waveform monitor is also used to determine the luminance (brightness and contrast) of a color signal and strength of the video level of that signal. The monitor measures the signal balance of color cameras, videotape recorders, and other electronic units.

Weather Channel (WC), The   a BASIC CABLE channel which is television's only 24-hour weather network. The service has developed a system for simultaneously transmitting different local weather reports to some 800 zones around the country.

Westar satellites   See HUGHES COMMUNICATIONS.

Westinghouse Broadcasting Company   See GROUP W.

White House Correspondents Association (WHCA)   a nonprofit membership association consisting of newspaper, magazine, and radio and television reporters covering the White House.

wholesaler   a company or individual serving as a middleman between the PROGRAM SUPPLIER and the retail video store. The term is used interchangeably with "distributor." The wholesaler buys home-video titles from the producer or program supplier and sells or rents them to a video store or school, college, or library. Wholesalers retain all or most of the income for their services, depending on the arrangement with the program supplier. They usually operate on a 12 to 16 percent MARGIN with little or no type of geographic or other type of retail exclusivity accorded to them by the program suppliers.

Home video distributors differ from television distributors, who only license programs. In home video, the ownership of the physical product (the videocassette or videodisc) actually changes hands in the TWO-STEP DISTRIBUTION process. Most large wholesalers belong to the NATIONAL ASSOCIATION OF VIDEO DISTRIBUTORS (NAVD). (See also RACK JOBBER.)

window dub   See DUBBING.

windows   the periods of availability in the various media for theatrical films and some MADE-FOR-TV MOVIES. The term came into use in the late 1970s with the advent of home video and the growth of cable television. Theatrical films are usually released to home video after their theatrical run, followed by sales to PAY PER VIEW (PPV) systems, and then to PAY (PREMIUM) CABLE SERVICE networks. Only then are the films licensed to the television networks, and finally to individual stations. The amount of time allowed for each step in the sequence varies from film to film, but each period is referred to as a "window" in the distribution cycle.

The sequence of the release of motion pictures is dependent on the particular deal and specific film, however, and the length of the windows can vary in any of the stages.

wipe a sophisticated form of a DISSOLVE that gradually replaces one moving image with another. The effect is of one picture "wiping out" the preceding one. The wipe was made possible in the late 1960s with the introduction of the SPECIAL-EFFECTS GENERATOR (SEG). A video line is manipulated to move across the screen from left to right or right to left or to move up and down or down and up. As the line is manipulated, the image on one camera is moved aside. (See also DIGITAL VIDEO EFFECTS.)

wireless cable See MULTICHANNEL MULTIPOINT DISTRIBUTION SERVICE (MMDS).

Wireless Cable Association (WCA) a nonprofit membership organization consisting of individuals and corporations involved in what is sometimes called "over-the-air cable" operations. Wireless cable seeks to offer the same kind of programming that exists on BASIC CABLE and PAY (PREMIUM) CABLE channels, by using MULTICHANNEL MULTIPOINT DISTRIBUTION SERVICE (MMDS) and INSTRUCTIONAL TELEVISION FIXED SERVICE (ITFS) systems.

Women in Broadcast Technology (WBT) a small nonprofit association that operates as a networking group for women who are employed as technical and engineering personnel in broadcasting. The organization holds workshops and publishes information about job opportunities and technical matters.

Women in Cable (WIC) a membership association of professional women in the cable industry. The nonprofit organization commissions studies about working conditions and opportunities for women in the field.

Women in Communications (WIC) a nonprofit association that is also a professional society of women in all fields of communication including journalism, radio, television, and cable. WIC supervises local chapters in more than 80 cities and 100 student chapters at schools and colleges throughout the nation.

workprint film footage that is assembled into a general order and sequence without sound or any optical effects (such as DISSOLVES or SUPERIMPOSITIONS). Sometimes known as a rough cut, this print usually undergoes many editing changes before the separate audio, video, and optical portions are sent back to the film laboratory to be combined into a composite or ANSWER PRINT.

World Administrative Radio Conference (WARC) See INTERNATIONAL TELECOMMUNICATIONS UNION (ITU).

World Communication Association (WCA) a nonprofit international organization that encourages all aspects of communication and conducts training programs, gives awards, and holds competitions and a biannual conference. The membership of the organization consists of researchers, teachers, and students located throughout the world.

World Institute of Black Communications (WIBC) an organization that helps increase the opportunities for blacks in the communications field and informs advertising agencies about the black consumer market. It compiles DEMOGRAPHIC research on blacks and maintains an extensive library.

World Standard Teletext (WST) a British-developed teletext system. Based on the successful experiences with both the ORACLE and CEEFAX one-way teletext operations in the United Kingdom, World Standard Teletext has been aggressively promoted as the standard for all teletext systems. Because the FEDERAL COMMUNICATIONS COMMISSION (FCC) has declined to establish a teletext standard in the United States, the marketplace will eventually establish a *de facto* standard. (See also ELECTRA and NORTH AMERICAN BROADCAST TELETEXT STANDARD [NABTS].)

Worldnet a global satellite service of the UNITED STATES INFORMATION AGENCY (USIA). It is designed to link newsmakers and experts in Washington with foreign journalists abroad through interactive press conferences. The service transmits via satellite from a studio in Washington D.C. It offers interviews with important members of the U.S. government to journalists at American embassies in more than 100 cities on six continents. Worldnet also transmits CABLE SATELLITE PUBLIC AFFAIRS (C-SPAN) programs.

WORM a type of CD-ROM videodisc that allows the user to create data. The initials stand for "write once, read many times." Using a personal computer (PC), an operator can input digital data such as office records or archival information onto a blank disc. Once entered, the data cannot be changed. This limits flexibility but ensures a permanent record for the data, which can be rapidly and easily summoned at any time by using the CD-ROM device.

wrap the finish of a show. The terms is used to describe the conclusion of the day's work or, more often, the ending of production for the entire project. The term was borrowed from the idea of wrapping up an object to encase and secure it for transportation, in this case, to put a program "in the can." The phrase "it's a wrap" means that the production is completed and everyone can go to the traditional wrap party.

wraparound programming programming that is sometimes scheduled to appear before and after a major nonfiction program that addresses the same theme or topic. Like the bread in a sandwich, the shows wrap around the meat of the main program. In a public affairs evening on PUBLIC TELEVISION (PTV) stations, a half-hour local lecture on drugs might be followed by an hour-long national DOCUMENTARY from the PUBLIC BROADCASTING SERVICE (PBS), followed by a half-hour local panel discussion about how the problem affects the community. The two half-hour programs before and after the documentary are called wraparound programs. If there is only one show and it comes after the documentary, it is called a FOLLOW-UP PROGRAM.

Z

zapping the practice of rapidly switching channels on a television set. Zapping is the physical manifestation of the process of GRAZING. Using remote control units, viewers can switch among the many channels available on cable systems, and the practice of zapping rapidly from one channel to another until a program catches the viewer's attention has become routine. Others zap the channels or turn off the sound during the COMMERCIALS in the program they are watching. The A. C. NIELSEN company, in fact, defines the term as the practice of eliminating commercials on a videocassette, but most industry practitioners use ZIPPING to describe that process.

Zapple doctrine a FEDERAL COMMUNICATIONS COMMISSION (FCC) rule that applies the FAIRNESS DOCTRINE to political broadcasting and is also an adjunct to the EQUAL TIME (OPPORTUNITY) RULES. Named after congressional aide Nicholas Zapple, the doctrine specifies that if a station sells or gives time to supporters of a candidate, it must give an equal opportunity to supporters of an opposing candidate. It covers situations where the candidates themselves do not appear, but their campaign is represented by spokespersons.

zipping the process of using a remote control device to fast-forward a prerecorded videocassette to a new segment, scene, or spot. It is similar to ZAPPING. The viewer can bypass a boring section of a movie or find "the good parts" to show someone. It sometimes becomes a form of GRAZING, although it is necessarily confined to one program or film at a time, rather than many channels and programs, as is the case with ZAPPING.

zoom lens a convenient and versatile television device that is actually a combination of lenses. It has a variable (rather than a fixed) FOCAL LENGTH and can operate in a range from a wide angle to a telephoto lens without losing focus. The lens can be used to change angles between shots or to zoom in on or away from a subject during a shot. The term probably originated in the early days of flight when it was used to describe the sound of a plane swooping down and in and up and out at an air show.

ABBREVIATIONS AND ACRONYMS

A&E   Arts and Entertainment Network
AA   average audience
AAF   American Advertising Federation
AAMSL   American Association of Media Specialists and Librarians
AAVT   Association of Audio-Visual Technicians
ABC   American Broadcasting Company
ABU   Asian Broadcasting Union
AC   alternating current
ACA   Association for Communication Administration
ACATS   Advisory Committee on Advanced Television Service
ACE   Award of Cable Excellence
ACEJMC   Accrediting Council on Education in Journalism and Mass Com-
           munications
ACI   Advertising Council Inc.
ACT   Action for Children's Television
ACTAT   Association of Cinematographers, TV, and Allied Technicians
ACTRA   Alliance of Canadian Cinema, Television, and Radio Artists
ACTV   advanced compatible television
ADI   area of dominant influence
ADR   alternative dispute resolution
ADTV   advanced definition television
AEA   American Electronics Association
AECT   Association for Educational Communications and Technology
AEJMC   Association for Education in Journalism and Mass Communication
AERho   Alpha Epsilon Rho
AES   Audio Engineering Society
AETT   Association for Educational and Training Technology
AFA   Advertising Federation of America
AFBA   Armed Forces Broadcasters Association
AFFFT   Academy of Family Films and Family Television
AFI   American Film Institute
AFM   American Federation of Musicians of the U.S. and Canada
AFMA   American Film Marketing Association
AFRTS   Armed Forces Radio and Television Service
AFT   automatic fine tuning
AFTRA   American Federation of Television and Radio Artists
AFVA   American Film and Video Association
AGC   automatic gain control
AICE   Association of Independent Commercial Editors
AICP   Association of Independent Commercial Producers
AIM   Accuracy in Media
AIME   Association for Information Media and Equipment
AIT   Agency for Instructional Technology

AIVF   Association of Independent Video and Film Makers
ALA   American Library Association
ALS   Adult Learning Services
AM   amplitude modulation
AMMI   American Museum of the Moving Image
AMST   Association of Maximum Service Telecasters
AMTEC   Association for Media and Technology in Education in Canada
ANA   Association of National Advertisers
ANI   automatic number identification
APB   Association for Public Broadcasting
APTS   America's Public Television Stations
AQH   quarter-hour audience
ARB   American Research Bureau Inc.
ARF   Advertising Research Foundation
ASA   American Sportscasters Association
ASCAP   American Society of Composers, Authors, and Publishers
ASHET   American Society for Healthcare, Education, and Training
ASTA   Advertiser Syndicated Television Association
ASTD   American Society for Training and Development
ASTVC   American Society for TV Cameramen
ATAS   Academy of Television Arts and Sciences
ATC   American Television and Communications Corporation
ATEL   Advanced Television Evaluation Laboratory
ATSC   Advanced Television Systems Committee
ATV   advanced television
AVC   Association of Visual Communicators
AVMA   Audio-Visual Management Association
AWRT   American Women in Radio and Television
AT&T   American Telephone and Telegraph

BAFTA   British Academy of Film and Television Arts
BAIT   Black Awareness in Television
BAR   Broadcast Advertising Reports
BBC   British Broadcasting Corporation
BDA   Broadcast Designers Association
BEA   Broadcast Education Association
BFMA   Broadcast Financial Management Association
BMI   Broadcast Music Inc.
BPME   Broadcast Promotion and Marketing Executives Inc.
BRC   Broadcast Rating Council
BTA   best time available
BTR   Broadcast Traffic and Residuals
BUFVC   British Universities Film and Video Council Ltd.

C-Span   Cable Satellite Public Affairs Network
CAB   Cabletelevision Advertising Bureau
CAB   Canadian Association of Broadcasters
CAI   computer-assisted instruction
CAMERA   Canadian Association of Motion Picture and Electronic Recording Artists
CAR   community antenna relay system
CATA   Community Antenna Television Association

CATV   community antenna television
CATV   Cable television
CAV   constant angular velocity
CBA   Commonwealth Broadcasting Association
CBA   Community Broadcasters Association
CBC   Canadian Broadcasting Corporation
CBT   computer-based training
CCD   charge-coupled devices
CCSN   Community College Satellite Network
CCTA   Canadian Cable Television Association
CCTV   closed-circuit television
CCUMC   Consortium of College and University Media Centers
CD   compact disc
CED   capacitance electronic disc
CEG   Consumer Electronics Group
CEN   Central Educational Network
CFEG   Canadian Film Editors Guild
CG   character generator
CINE   Council on International Nontheatrical Events
CIT   Commission on Instructional Technology
CLV   constant linear velocity
CMSA   Consolidated Metropolitan Statistical Area
CNBC   Consumer News and Business Channel
CNN   Cable News Network
CNTV   Center for New Television
COLTAM   Committee on Local Television Audience Measurement
Comsat   Communications Satellite Corporation
CONUS   continental U.S.
CP   construction permit
CPB   Corporation for Public Broadcasting
CPE   Columbia Pictures Entertainment
CPM   cost per thousand
CPP   cost per point
CPRP   cost per rating point
CRT   cathode ray tube
CRT   Copyright Royalty Tribunal
CRTC   Canadian Radio-Television and Telecommunications Commission
CSG   community service grants
CTAM   Cable Television Administration and Marketing Society
CTW   Children's Television Workshop
CU   closeup
CVC   Cablevision Systems Corporation

DA   distribution amplifier
DAT   digital audio tape
dB   decibel
DBS   direct broadcast satellite
DC   direct current
DEMM   Division of Educational Media Management
DGA   Directors Guild of America
DGC   Directors Guild of Canada
DMA   designated market area

DOC   Department of Communications
DOT   Division of Telecommunications
DSMS   Division of School Media Specialists
DVE   digital video effects
DVS   Descriptive Video Service

EBS   emergency broadcast system
EBU   European Broadcasting Union
ECU   extreme closeup
EDI   electronic data interchange
EDTV   enhanced definition television
EEN   Eastern Educational Network
EEO   equal employment opportunity
EFLA   Educational Film Library Association
EIA   Electronic Industries Association
EIAJ   Electronic Industries Association of Japan
EMRC   Electronic Media Rating Council
ERA   Electronics Representatives Association
ERP   effective radiated power
ETA-I   Electronic Technicians Association, International
ETV   educational television

FAB   Film Advisory Board
FAIR   fairness and accuracy in reporting
fc   footcandle
FCC   federal communications commission
FISR   financial interest-syndication rules
FM   frequency modulation
4As   American Association of Advertising Agencies
FTC   federal trade commission

GHz   gigahertz
GRP   gross rating point

HBI   Hubbard Broadcasting Inc.
HBO   Home Box Office
HDTV   high definition television
HeSCA   Health Sciences Communication Association
HSN   Home Shopping Networks I and II
HUT   households using television
Hz   Hertz

IAA   International Advertising Association
IABC   International Association of Business Communicators
IASUS   International Association of Satellite Users and Suppliers
IATSE   International Alliance of Theatrical Stage Employees
IBA   Independent Broadcasting Authority
IBEW   International Brotherhood of Electrical Workers
ICEM   International Council for Educational Media
ICIA   International Communications Industry Association
IDTV   improved definition television
IEEE   Institute of Electrical and Electronics Engineers

IGITT   Interagency Group for Interactive Training Technologies
IIC   International Institute of Communications
IMA   Interactive Multimedia Association
Intelsat   International Telecommunications Satellite Organization
INTV   Association of Independent Television Stations
IO   image orthicon tube
IPS   Interregional Program Service
IRTS   International Radio and Television Society
ISCET   International Society of Certified Electronics Technicians
ISCS   International Society of Communications Specialists
ISV   International Society of Videographers
ITA   International Tape/Disc Association
ITCA   International Teleconferencing Association
ITFS   instructional television fixed service
ITS   International Teleproduction Society
ITU   International Telecommunications Union
ITV   instructional television
ITVA   International Television Association
IUC   International University Consortium
IVIA   Interactive Video Industry Association

JESCOM   Jesuits in Communication in the U.S.

K   Kelvin
KHz   kilohertz

LITA   Library and Information Technology Association
LOP   least objectionable program
LPTV   low power television
LS   long shot
LUC   lowest unit charges
LUD   limited use discount
LV   laser videodisc
LVR   longitudinal videotape recording

MAP   minimum advertised price
MATV   master antenna television
MB   Museum of Broadcasting
MBC   Museum of Broadcasting Communications
MCTV   multichannel television
MEU   Mind Extension University
MHz   megahertz
MLCA   Multilingual Communications Association
MMDS   multichannel multipoint distribution service
MPAA   Motion Picture Association of America
MRS   minimum reporting standards
MS   medium shot
MSA   Metropolitan Statistical Area
MSO   multiple system operator

NAATA   National Asian American Telecommunication Association
NAB   National Association of Broadcasters

NABET   National Association of Broadcast Employees and Technicians
NABOB   National Association of Black-Owned Broadcasters
NABTS   North American Broadcast Teletext Standards
NACB   National Association of College Broadcasters
NACP   National Academy of Cable Programming
NAMB   National Association of Media Brokers
NAMID   National Moving Image Data Base
NAPTS   National Association of Public Television Stations
NARDA   National Association of Retail Dealers of America
NARM   National Association of Recording Merchandisers
NARMC   National Association of Regional Media Centers
NARTE   National Association of Radio and Telecommunications Engineers
NASTEMP   National Association of State Educational Media Professionals
NATAS   National Academy of Television Arts and Sciences
NATOA   National Association of Telecommunications Officers and Advisors
NAVC   National Audio-Visual Center
NAVD   National Association of Video Distributors
NBC   National Broadcasting Company
NBMC   National Black Media Coalition
NBPC   National Black Programming Consortium
NBPC   National University Consortium for Telecommunications in Teaching
NCI   National Captioning Institute
NCTA   National Cable Television Association
NCTC   National Cable Television Cooperative Inc.
NCTI   National Cable Television Institute
NEDA   National Electronic Distributors Association
NESSDA   National Electronics Sales and Service Dealer Association
NFB   National Film Board of Canada
NFLCP   National Federation of Local Cable Programmers
NHI   Nielsen Homevideo Index
NHK   Japan Broadcasting Corporation
NICEM   National Information Center for Educational Media
NPR   National Public Radio
NRB   National Religious Broadcasters
NSI   Nielsen Station Index
NSS   Nielsen Syndicated Service
NTA   National Translator Association
NTI   Nielsen Television Index
NTIA   National Telecommunications and Information Administration
NTU   National Technological University
NWICO   New World Information and Communication Order

OFS   operations fixed service
OTO   one-time-only

PARA   Professional Audio/Video Retailers Association
PBFA   Public Broadcasting Financing Act of 1975
PBS   Public Broadcasting Service
PC   personal computer
PEG   public access, educational
PFVEA   Professional Film and Video Equipment Association
PMN   Pacific Mountain Network
PMSA   Primary Metropolitan Statistical Area

POP   point-of-purchase
PP   pocketpiece
PPT   pay-per-transaction
PPV   pay per view
PRCA   Pacific Rim Coproduction Association
Prestel   press and tell
PSA   public service announcement
PTAR   prime time access rule
PTC   Pacific Telecommunications Council
PTFMA   Public Telecommunications Financial Management Association
PTV   public television
PUT   people using television

quad   quadruplex

RCA   Radio Corporation of America
reps   station representatives
ROS   run-of-schedule
RTNDA   Radio Television News Directors Association
RTRC   Radio and Television Research Council

S/N   signal-to-noise ratio
SAG   Screen Actors Guild
SAP   separate audio program
SBCA   Satellite Broadcasting and Communications Association
SBE   Society of Broadcast Engineers
SCA   Speech Communication Association
SCA   Subsidiary Communication Authorization Service
SCJ   Society for Collegiate Journalists
SCTE   Society of Cable Television Engineers
SECA   Southern Educational Communications Association
SEG   Screen Extras Guild
SEG   special-effects generator
SERC   Satellite Educational Resources Consortium
SEV   special event video
SHDTV   simulcast high definition television
SHL   studio-headend link
SI   special interest
SIA   Storage Instantaneous Audimeter
SIN   Spanish International Network
SIVA   Special Interest Video Association
SMATV   satellite master antenna television
SMPTE   Society of Motion Picture and Television Engineers
SNG   satellite news gathering
SRDS   Standard Rate and Data Service
SSPI   Society of Satellite Professionals International
STA   special temporary authorization
STL   studio-transmitter link
STV   subscription television
sync   synchronization

TAR   total audience rating
TBC   time base corrector

TCA   Television Critics Association
TCI   Telecommunications Inc.
TERC   Technical Education Research Center
TLC   The Learning Channel
TMC   The Movie Channel
TRAC   Telecommunications Research and Action Center
TvB   Television Bureau of Advertising
TVRO   television receive only

UA   United Artists Entertainment
UCIDT   University Consortium for Instructional Development and Technology
UFVA   University Film and Video Association
UNCA   United Nations Correspondents Association
USIA   United States Information Agency
USSB   United States Satellite Broadcasting

VCR   videocassette recorders
VF   video floppy
VHD   video high density
VHDTV   very high definition TV
VIA   Videotext Industry Association
VRA   Video Retailers Association
VSDA   Video Software Dealers Association
VTR   videotape recorder

WARC   World Administrative Radio Conference
WBT   Women in Broadcast Technology
WCA   Wireless Cable Association
WCA   World Communication Association
WHCA   White House Correspondents Association
WIBC   World Institute of Black Communications
WIC   Women in Cable
WIC   Women in Communications
WST   World Standard Teletext